COLLEGE PHYSICS

By

D.K. Jha

DISCOVERY PUBLISHING HOUSE PVT. LTD.
NEW DELHI-110 002

First Published-2008
Reprint: 2011
ISBN 978-81-8356-394-9

Published by:
DISCOVERY PUBLISHING HOUSE PVT. LTD.
4831/24, Ansari Road, Prahlad Street,
Darya Ganj, New Delhi-110002 (India)
Phone: 23279245 • Fax: 91-11-23253475
E-mail: dphbooks@rediffmail.com
dphtemp@indiatimes.com
Website: www.discoverypublishinghouse.com

Printed at:
Dynamic Printers, Delhi

Contents

Preface

Physics is a science that seeks to understand very basic concepts, such as force, energy, mass and charge. In other words, it is the science of matter and its motion. More precisely, it is the general analysis of nature, conducted in order to understand, how the world around us behaves. In one form or the other, physics is one of the oldest academic disciplines, perhaps the oldest through inclusion of astronomy in it. Physics is both significant and influential, in part, because advances in its understanding have often translated into new technologies, and also because new ideas in physics often resonate with other sciences and mathematics and philosophy.

In modern science, when we talk of physics, it is useful to further subdivide the subject into various disciplines, including astrophysics, atomic physics, molecular physics, biophysics, solid state physics, optical and laser physics, fluid and plasma physics, nuclear physics and particle physics, etc. Progress in physics frequently comes about, when experimentalists make a discovery that existing theories cannot suffice further, necessitating the formulation of new theories. Likewise ideas, arising from one theory, often inspire new experiments. Research in physics is continually progressing on larger fronts.

Realising the need of a specialised book on 'College Physics' that covers all the aspects of the subject, we have made this effort. Hopefully, with this book the students and general readers would be benefited, equally.

Enlightening feedback is always solicited

Introduction

Physics, in everyday terms, is the science of matter and its motion. It is the science that seeks to understand very basic concepts such as force, energy, mass and charge. More completely, it is the general analysis of nature, conducted in order to understand how the world around us behaves.

In one form or another, physics is one of the oldest academic disciplines, perhaps the oldest through its inclusion of astronomy. Over the last two millennia, physics has sometimes been synonymous with philosophy, chemistry and certain branches of mathematics and biology but it emerged as a modern science in the 16th century. Physics is now generally distinct from these other disciplines, even though its boundaries remain difficult to define rigorously.

Physics is significant and influential, in part because advances in its understanding have often translated into new technologies, but also because new ideas in physics often resonate with the other sciences, mathematics and philosophy. For example, advances in the understanding of electromagnetism led directly to the development of new products that have transformed society (including television, computers and domestic appliances); advances in thermodynamics led to the development of motorised transport; and advances in mechanics inspired the development of the calculus, quantum chemistry, and the use of instruments like the electron microscope in microbiology.

Today, physics is both a broad and deep subject that, in practical terms, can be split into several subfields. It can also be divided into two conceptually different branches: theoretical and experimental physics; the former dealing with the development of new theories, whilst the latter deals with the experimental testing of these new, or existing, theories. Despite many important discoveries during the last four centuries, many significant questions about nature still remain unanswered, and many areas of the subject are still highly active.

Scope and Goals

Physics is the discipline devoted to understanding nature in a very general sense: the fundamental characteristic of physics is that it aims to gain knowledge, and hopefully understanding of the general properties of the world around us. As an example, we can consider asking the following question on the nature of the Universe itself: how many dimensions do we need? Given that we know the universe to consist of four dimensions (three space dimensions, and one time dimension), we can also ask why the universe picked those particular numbers? Why not have four space dimensions?

The fact that a choice was made out of a possibility of many means that questions like these fall under the scope of physics. Other general properties of nature include the existence of mass (as in Newton's laws of motion), charge (as in Maxwell's equations), and spin (in Quantum mechanics), amongst others.

However, whilst physics studies the general properties of nature, it will often also study the properties of certain objects within nature. Thus, it is also physics whose job it is to describe what happens to, for example, planets whose motion is affected by nearby stars. Generally, the study of the specific objects in nature are shared between the three sciences: biology is roughly responsible for the living organisms, chemistry for the study of the elements and molecules, and physics is given responsibility over all that remains. The fact that physics is delegated all objects besides those covered by biology and chemistry means that it is responsible for the study of a wide range objects and phenomena, from the smallest subatomic particles, to the largest galaxies. Included in this are the very most basic objects from which all other things are composed of, and therefore physics is sometimes said to be the "fundamental science".

Generalities aside, physics aims to describe the various phenomena in nature in terms of simpler phenomena: that is, to find the mechanisms for why nature behaves the way it does. Thus, physics aims to both connect the things we see around us to a root cause, and then to try to connect these root causes together in the hope of finding an ultimate reason for why nature is as it is. For example, the ancient Chinese observed that certain rocks (lodestone) were attracted to one another by some invisible force. This effect was later called *magnetism*, and was first rigorously studied in the 17th century. A little earlier than the Chinese, the ancient Greeks knew of other objects (amber) that when rubbed with fur would cause a similar invisible attraction between the two. This was also first studied rigorously in the 17th century, and came to be called *electricity*. Thus, physics had come to understand two observations of nature in terms of some root cause (electricity and magnetism). However, further work in the 19th century revealed that these two forces were just two different aspects of one force — electromagnetism. This process of "unifying" forces continues today.

Physics Uses and Scientific Methods

Physics uses the scientific method. That is, that the sole test of the validity of a physical theory be comparison with observation. Experiments and observations are to be collected and matched with the predictions of theories, thus verifying or falsifying the theory.

Those theories which are very well supported by data and which are especially simple and general have been called *scientific laws*. Of course, all theories, including those called *laws*, can also be replaced by more accurate and more general statements, if a disagreement of theory with observed data were to be found.

Data Collection and Theory Development

There are many approaches to studying physics, and many different kinds of activities in physics. There are two main types of activities in physics; the collection of data and the development of theories.

The data in some subfields of physics is amenable to experiment. For example, condensed matter physics and nuclear physics benefit from the ability to perform experiments. Sometimes experiments are done to explore nature, and in other cases experiments are performed to produce data to compare with the predictions of theories.

Some other fields in physics like astrophysics and geophysics are primarily observational sciences because most their data has to be collected passively instead of through experimentation. Nevertheless, observational programmes in these fields uses many of the same tools and technology that are used in the experimental subfields of physics. The accumulated body of knowledge in some area of physics through experiment and observation is known as *phenomenology*.

Theoretical physics often uses quantitative approaches to develop the theories that attempt to explain the data. In this way, theoretical physics often relies heavily on tools from mathematics and computational technologies (particularly in the subfield known as *computational physics*). Theoretical physics often involves creating quantitative predictions of physical theories, and comparing these predictions quantitatively with data. Theoretical physics sometimes creates models of physical systems before data are available to test and validate these models.

These two main activities in physics, data collection and theory production and testing, draw on many different skills. This has lead to a lot of specialisation in physics, and the introduction, development and use of tools from other fields. For example, theoretical physicists apply mathematics and numerical analysis and statistics and probability and computers and computer software in their work. Experimental physicists develop instruments and techniques for collecting data, drawing on engineering and computer technology and many other fields of technology. Often the tools from these other areas are not quite appropriate for the needs of physics, and need to be adapted or more advanced versions have to be produced.

The culture of physics research differs from the other sciences in the separation of theory from data collection through experiment and observation. Since the 20th century, most individual physicists have specialised in either theoretical physics or experimental physics. The great Italian physicist Enrico Fermi (1901-1954), who made fundamental contributions to both theory and experimentation in nuclear physics, was a notable exception. In contrast, almost all the successful theorists in biology and chemistry (e.g. American quantum chemist and biochemist Linus Pauling) have also been experimentalists, though this is changing as of late.

Although theory and experiment are usually performed by separate groups, they are strongly dependent on each other. Progress in physics frequently comes about when experimentalists make a discovery that existing theories cannot account for, necessitating the formulation of new theories. Likewise, ideas arising from theory often inspire new experiments.

In the absence of experiment, theoretical research can go in the wrong direction; this is one of the criticisms that has been levelled against M-theory, a popular theory in high-energy physics for which no practical experimental test has ever been devised.

Physics being Quantitative

Physics is more quantitative than most other sciences. That is, many of the observations experimental results in physics are numerical measurements. Most of the theories in physics use mathematics to express their principles. Most of the predictions from these theories are numerical. This is because of the areas which physics has addressed are more amenable to quantitative approaches than other areas. Physical definitions, models and theories can often be expressed using mathematical relations, as early as 1638, when Galileo published the law of falling bodies in his *Two New Sciences*.

A key difference between physics and mathematics is that because physics is ultimately concerned with descriptions of the material world, it tests its theories by comparing the predictions of its theories with data from observations or experiments, whereas mathematics is concerned with abstract logical patterns not limited by those observed in the real world (because the real world is limited in the number of dimensions and in many other ways it does not have to correspond to richer mathematical structures).

The distinction, however, is not always clear-cut. There is a large area of research intermediate between physics and mathematics, known as *mathematical physics*.

Relation to Mathematics and Other Sciences

Physics relies on mathematics to provide the logical framework in which physical laws can be precisely formulated and their predictions quantified. Physical definitions, models and theories can be succinctly expressed using mathematical relations.

Whenever analytic solutions are not feasible, numerical analysis and simulations can be utilised. Thus, scientific computation is an integral part of physics, and the field of computational physics is an active area of research.

Beyond the known universe, the field of theoretical physics also deals with hypothetical issues, such as parallel universes, a multiverse, or whether the universe could have expanded as predominantly anti-matter rather than matter.

In the *Assayer* (1622), Galileo noted that mathematics is the language in which nature expresses laws, to be discovered by physicists. Physics is also intimately related to many other sciences, as well as applied fields like engineering and medicine. The principles of physics find applications throughout the other natural sciences as they depend on the interactions of the fundamental constituents of the natural world. Some of the phenomena studied in physics, such as the phenomenon of conservation of energy, are common to all material systems. These are often referred to as laws of physics. Others, such as superconductivity, stem from these laws, but are not laws themselves because they only appear in some systems. Physics is often said to be the "fundamental science" (chemistry is sometimes included), because each of the other disciplines (biology, chemistry, geology, material science, engineering, medicine, etc.) deals with particular types of material systems that obey the laws of physics. For example, chemistry is the science of collections of matter (such as gases and liquids formed of atoms and molecules) and the processes known as chemical reactions that result in the change of chemical substances. The structure, reactivity, and properties of a chemical compound are determined by the properties of the underlying molecules, which can be described by areas of physics such as quantum mechanics (called in this case quantum chemistry), thermodynamics, and electromagnetism.

Philosophical Implications

Physics in many ways stemmed from ancient Greek philosophy. From Thales' first attempt to characterise matter, to Democritus' deduction that matter ought to reduce to an invariant state, to the Ptolemaic Astronomy of a crystalline firmament upon which the stars rested, our view of the universe seemed static. By the twentieth century, this picture became less certain, and now a static universe is only one possibility in an array of possible universes.

Aristotle's early observations in natural history, and natural philosophy usually did not involve much fact checking or detailed observation, which allowed errors to come to rest in our knowledge of the world. When closer investigation overturned this picture of the world, philosophers came to study other possible forms of reasoning. The use of a priori reasoning found a natural place in scientific method as well as the use of experiments and a posteriori reasoning came to be used in Bayesian inference. By the 19th century, physics was realised as a positive science and a distinct discipline separate from philosophy and the other sciences.

Study of the philosophical issues surrounding physics, the philosophy of physics can be encapsulated as empiricism, naturalism, and for some, realism. The mathematical physicist Roger Penrose has been called a *Platonist* by Stephen Hawking, while Penrose continues to eschew quantum mechanics as a final theory about reality.

Orsted (1811) noted that physicists readily make deductions about nature, based on their closer familiarity with experiments about nature, whereas the mathematicians and philosophers must make do with fewer positive statements about nature.

There are certain statements, such as Newton's Third Law of Motion, that can be generalised into the Principle of Equivalence. This principle is the logical basis for general relativity, whose solutions give metrics for spacetime. The success of general relativity influenced Einstein to eschew quantum theory, to which he made seminal contributions, and to eventually believe that all physical theory ought to be independent of observation.

Topic Outlines

This list of basic physics topics covers much of physics, the science concerned with the discovery and understanding of the fundamental laws which govern matter, energy, space, and time. Physics deals with the elementary constituents of the universe and their interactions, as well as the analysis of systems best understood in terms of these fundamental principles. Because physics treats the core workings of the universe, including the quantum mechanical details which underpin all atomic interactions, it can be thought of as the foundational science, upon which stands "the central science" of chemistry, the earth sciences, biological sciences, and social sciences. Discoveries in basic physics have important ramifications for all of science.

Essence of Physics

Physics started with a philosophical commitment to simplicity. It should not be considered a difficult subject (although it is deep); one can learn classical physics on a playground, which describes the motion of balls, swings, slides and merry-go-rounds.

Classical Mechanics

Classical mechanics is used for describing the motion of macroscopic objects, from projectiles to parts of machinery, as well as astronomical objects, such as spacecraft, planets, stars, and galaxies. It produces very accurate results within these domains, and is one of the oldest and largest subjects in science and technology.

Besides this, many related specialities exist, dealing with gases, liquids, and solids, and so on. Classical mechanics is enhanced by special relativity for objects moving with high velocity, approaching the speed of light; general relativity is employed to handle gravitation at a deeper level; and quantum mechanics handles the wave-particle duality of atoms and molecules.

In physics, classical mechanics is one of the two major subfields of study in the science of mechanics, which is concerned with the set of physical laws governing and mathematically describing the motions of bodies and aggregates of bodies. The other subfield is quantum mechanics.

The term classical mechanics was coined in the early 20th century to describe the system of mathematical physics begun by Isaac Newton and many contemporary 17th century workers, building upon the earlier astronomical theories of Johannes Kepler, which in turn were based on the precise observations of Tycho Brahe and the studies of terrestrial projectile motion of Galileo, but before the development of quantum physics and relativity. Therefore, some sources exclude so-called "relativistic physics" from that category. However, a number of modern sources do include Einstein's mechanics, which in their view represents classical mechanics in its most developed and most accurate form. The initial stage in the development of classical mechanics is often referred to as Newtonian mechanics, and is associated with the physical concepts employed by and the mathematical methods invented by Newton himself, in parallel with Leibniz, and others. More abstract and general methods include Lagrangian mechanics and Hamiltonian mechanics. Much of the content of classical mechanics was created in the 18th and 19th centuries and extends considerably beyond (particularly in its use of analytical mathematics) the work of Newton.

Electromagnetism

Electromagnetism is the physics of the electromagnetic field: a field which exerts a force on particles that possess the property of electric charge, and is in turn affected by the presence and motion of those particles.

A changing magnetic field produces an electric field (this is the phenomenon of electromagnetic induction, the basis of operation for electrical generators, induction motors, and transformers). Similarly, a changing electric field generates a magnetic field. Because of this interdependence of the electric and magnetic fields, it makes sense to consider them as a single coherent entity — the electromagnetic field.

The magnetic field is produced by the motion of electric charges, i.e., electric current. The magnetic field causes the magnetic force associated with magnets.

The theoretical implications of electromagnetism led to the development of special relativity by Albert Einstein in 1905.

Theory of Relativity

The theory of relativity, or simply relativity, refers specifically to two theories of Albert Einstein: special relativity and general relativity. However, "relativity" can also refer to Galilean relativity.

The term "theory of relativity" was coined by Max Planck in 1908 to emphasise how special relativity (and later, general relativity) uses the principle of relativity.

Statistical Mechanics

Statistical mechanics is the application of probability theory, which includes mathematical tools for dealing with large populations, to the field of mechanics, which is concerned with the motion of particles or objects when subjected to a force. Statistical mechanics, sometimes called *statistical physics,* can be viewed as a subfield of physics and chemistry. Pioneers in establishing the field were Ludwig Boltzmann and Josiah Willard Gibbs.

It provides a framework for relating the microscopic properties of individual atoms and molecules to the macroscopic or bulk properties of materials that can be observed in everyday life, therefore explaining thermodynamics as a natural result of statistics and mechanics (classical and quantum) at the microscopic level. In particular, it can be used to calculate the thermodynamic properties of bulk materials from the spectroscopic data of individual molecules.

This ability to make macroscopic predictions based on microscopic properties is the main advantage of statistical mechanics over thermodynamics. Both theories are governed by the second law of thermodynamics through the medium of entropy. However, entropy in thermodynamics can only be known empirically, whereas in statistical mechanics, it is a function of the distribution of the system on its micro-states.

Quantum Mechanics

Quantum mechanics is the study of mechanical systems whose dimensions are close to the atomic scale, such as molecules, atoms, electrons, protons and other subatomic particles. Quantum mechanics is a fundamental branch of physics with wide applications. Quantum theory generalises classical mechanics to provide accurate descriptions for many previously unexplained phenomena such as black body radiation and stable electron orbits. The effects of quantum mechanics become evident at the atomic and subatomic level, and they are typically not observable on macroscopic scales. Superfluidity is one of the known exceptions to this rule.

The word "quantum" came from the Latin word which means "how great" or "how much." In quantum mechanics, it refers to a discrete unit that quantum theory assigns to certain physical quantities, such as the energy of an atom at rest. The discovery that waves have discrete energy packets (called quanta) that behave in a manner similar to particles led to the branch of physics that deals with atomic and subatomic systems which we today call quantum mechanics. It is the underlying mathematical framework of many fields of physics and chemistry, including condensed-matter physics, solid-state physics, atomic physics, molecular physics, computational chemistry, quantum chemistry, particle physics, and nuclear physics. The foundations of quantum mechanics were established during the first half of the twentieth century by Werner Heisenberg, Max Planck, Louis de Broglie, Albert Einstein, Niels Bohr, Erwin Schrodinger, Max Born, John von Neumann, Paul Dirac, Wolfgang Pauli and others. Some fundamental aspects of the theory are still actively studied.

Quantum mechanics is essential to understand the behaviour of systems at atomic length scales and smaller. For example, if classical mechanics governed the workings of an atom, electrons would rapidly travel towards and collide with the nucleus, making stable atoms impossible. However, in the natural world the electrons normally remain in an unknown orbital path around the nucleus, defying classical electromagnetism.

Quantum mechanics was initially developed to provide a better explanation of the atom, especially the spectra of light emitted by different atomic species. The quantum theory of the atom was developed as an explanation for the electron's staying in its orbital, which could not be explained by Newton's laws of motion and by Maxwell's laws of classical electromagnetism.

In the formalism of quantum mechanics, the state of a system at a given time is described by a complex wave function (sometimes referred to as orbitals in the case of atomic electrons), and more generally, elements of a complex vector space. This abstract mathematical object allows for the calculation of probabilities of outcomes of concrete experiments. For example, it allows one to compute the probability of finding an electron in a particular region around the nucleus at a particular time. Contrary to classical mechanics, one can never make simultaneous predictions of conjugate variables, such as position and momentum, with arbitrary accuracy. For instance, electrons may be considered to be located somewhere within a region of space, but with their exact positions being unknown. Contours of constant probability, often referred to as "clouds" may be drawn around the nucleus of an atom to conceptualise where the electron might be located with the most probability. Heisenberg's uncertainty principle quantifies the inability to precisely locate the particle.

The other exemplar that led to quantum mechanics was the study of electromagnetic waves such as light. When it was found in 1900 by Max Planck that the energy of waves could be described as consisting of small packets or quanta, Albert Einstein exploited this idea to show that an electromagnetic wave such as light could be described by a particle called the *photon* with a discrete energy dependent on its frequency. This led to a theory of unity between subatomic particles and electromagnetic waves called *wave–particle duality* in which particles and waves were neither one nor the other, but had certain properties of both. While quantum mechanics describes the world of the very small, it also is needed to explain certain "macroscopic quantum systems" such as superconductors and superfluids.

Broadly speaking, quantum mechanics incorporates four classes of phenomena that classical physics cannot account for: i) the quantisation (discretisation) of certain physical quantities, ii) wave-particle duality, iii) the uncertainty principle, and iv) quantum entanglement.

Branches of Physics

Astrophysics

Astrophysics is the branch of astronomy that deals with the physics of the universe, including the physical properties (luminosity, density, temperature, and chemical composition) of celestial objects such as stars, galaxies, and the interstellar medium, as well as their interactions. The study of cosmology is theoretical astrophysics at scales much larger than the size of particular gravitationally-bound objects in the universe.

Because astrophysics is a very broad subject, astrophysicists typically apply many disciplines of physics, including mechanics, electromagnetism, statistical mechanics, thermodynamics, quantum mechanics, relativity, nuclear and particle physics, and atomic and molecular physics. In practice, modern astronomical research involves a substantial amount of physics. The name of a university's department (astrophysics or astronomy) often has to do more with the department's history than with the contents of the programmes. Astrophysics can be studied at the bachelors, masters, and Ph.D. levels in aerospace engineering, physics, or astronomy departments at many universities.

Atomic Physics

Atomic physics (or atom physics) is the field of physics that studies atoms as an isolated system of electrons and an atomic nucleus. It is primarily concerned with the arrangement of electrons around the nucleus and the processes by which these arrangements change. This includes ions as well as neutral atoms and, unless otherwise stated, for the purposes of this discussion it should be assumed that the term atom includes ions.

The term "atomic physics" is often associated with nuclear power and nuclear bombs, due to the synonymous use of atomic and nuclear in standard English. However, physicists distinguish between atomic physics — which deals with the atom as a system comprising of a nucleus and electrons, and nuclear physics — which considers atomic nuclei alone.

As with many scientific fields, strict delineation can be highly contrived and atomic physics is often considered in the wider context of *atomic, molecular, and optical physics*. Physics research groups are usually so classified.

Biophysics

Biophysics (also biological physics) is an interdisciplinary science that employs and develops theories and methods of the physical sciences for the investigation of biological systems. Studies included under the umbrella of biophysics span all levels of biological organisation, from the molecular scale to whole organisms and ecosystems. Biophysical research shares significant overlap with biochemistry, nanotechnology, bioengineering and systems biology.

Molecular biophysics typically address biological questions that are similar to those in biochemistry and molecular biology, but the questions are approached quantitatively. Scientists in this field conduct research concerned with understanding the interactions between the various systems of a cell, including the interactions between DNA, RNA and protein biosynthesis, as well as how these interactions are regulated. A great variety of techniques are used to answer these questions.

Fluorescent imaging techniques, as well as electron microscopy, x-ray crystallography, NMR spectroscopy and atomic force microscopy (AFM) are often used to visualise structures of biological significance. Direct manipulation of molecules using optical tweezers or AFM can also be used to monitor biological events where forces and distances are at the nanoscale. Molecular biophysicists often consider complex biological events as systems of interacting units which can be understood through statistical mechanics, thermodynamics and chemical kinetics. By drawing knowledge and experimental techniques from a wide variety of disciplines, biophysicists are often able to directly observe, model or even manipulate the structures and interactions of individual molecules or complexes of molecules.

In addition to traditional (i.e. molecular) biophysical topics like structural biology or enzyme kinetics, modern biophysics encompasses an extraordinarily broad range of research. It is becoming increasingly common for biophysicists to apply the models and experimental techniques derived from physics, as well as mathematics and statistics, to larger systems such as tissues, organs, populations and ecosystems.

Chemical Physics

Chemical physics is a subdiscipline of chemistry and physics that investigates physicochemical phenomena using techniques from atomic and molecular physics and condensed matter physics; it is the branch of physics that studies chemical processes from the point of view of physics. While at the interface of physics and chemistry, chemical physics is distinct from physical chemistry in that it focuses more on the characteristic elements and theories of physics. Meanwhile, physical chemistry studies the physical nature of chemistry. Nonetheless, the distinction between the two fields is vague, and workers often practice in each field during the course of their research.

Chemical physicists commonly probe the structure and dynamics of ions, free radicals, polymers, clusters, and molecules. Areas of study include the quantum mechanical behaviour of chemical reactions, the process of solvation, inter and intra-molecular energy flow, and single entities such as quantum dots. Experimental chemical physicists use a variety of spectroscopic techniques to better understand hydrogen bonding, electron transfer, the formation and dissolution of chemical bonds, chemical reactions, and the formation of nanoparticles. Theoretical chemical physicists create simulations of the molecular processes probed in these experiments to both explain results and guide future investigations. The goals of chemical physics research include understanding chemical structures and reactions at the quantum mechanical level, elucidating the structure and reactivity of gas phase ions and radicals, and discovering accurate approximations to make the physics of chemical phenomena computationally accessible. Chemical physicists are looking for answers to such questions as:

- Can we experimentally test quantum mechanical predictions of the vibrations and rotations of simple molecules?
- Can we develop more accurate methods for calculating the electronic structure and properties of molecules?
- Can we understand chemical reactions from first principles?
- Why do quantum dots start blinking (in a pattern suggesting fractal kinetics) after absorbing photons of light?
- How do chemical reactions really take place?
- What is the step-by-step process that occurs when an isolated molecule becomes solvated?
- Can we use the properties of negative ions to determine molecular structures, understand the dynamics of chemical reactions, or explain photodissociation?
- Why does a stream of soft x-rays knock enough electrons out of the atoms in a xenon cluster to cause the cluster to explode?

Classical Physics

Classical physics is physics based on principles developed before the rise of general theory of relativity and quantum theory, usually including the special theory of relativity.

Among the branches of theory included in classical physics are:

- Classical mechanics:
  - — Newton's laws of motion;
  - — Classical Lagrangian and Hamiltonian formalisms.
- Classical electrodynamics (Maxwell's Equations).
- Classical thermodynamics.
- Special theory of relativity and General theory of relativity.
- Classical chaos theory and non-linear dynamics.

In contrast to classical physics, modern physics is a slightly looser term which may refer to just quantum physics or to 20th and 21st century physics in general and so always includes quantum theory and may.include relativity.

A physical system on the classical level is a physical system in which the laws of classical physics are valid. There are no restrictions on the application of classical principles, but, practically, the scale of classical physics is the level of isolated atoms and molecules on upwards, including the macroscopic and astronomical realm. Inside the atom and among atoms in a molecule, the laws of classical physics break down and generally do not provide a correct description.

Moreover, the classical theory of electromagnetic radiation is somewhat limited in its ability to provide correct descriptions, since light is inherently a quantum phenomenon. Unlike quantum physics, classical physics is generally characterised by the principle of complete determinism.

Mathematically, classical physics equations are ones in which Planck's constant does not appear. According to the correspondence principle and Ehrenfest's theorem as a system becomes larger or more massive. The classical dynamics tends to emerge, with some exceptions, such as superfluidity. This is why we can usually ignore quantum mechanics when dealing with everyday objects; instead the classical description will suffice. However, one of the most vigorous on-going fields of research in physics is classical-quantum correspondence. This field of research is concerned with the discovery of how the laws of quantum physics give rise to classical physics in the limit of the large scales of the classical level.

Condensed-matter Physics

Condensed-matter physics is the field of physics that deals with the macroscopic physical properties of matter. In particular, it is concerned with the "condensed" phases that appear whenever the number of constituents in a system is extremely large and the interactions between the constituents are strong. The most familiar examples of condensed phases are solids and liquids, which arise from the bonding and electromagnetic force between atoms. More exotic condensed phases include the superfluid and the Bose-Einstein condensate found in certain atomic systems at very low temperatures, the superconducting phase exhibited by conduction electrons in certain materials, and the ferromagnetic and anti-ferromagnetic phases of spins on atomic lattices. Condensed matter physics is that branch of physics which deals with the macroscopic physical properties of matter namely, solid and liquid, when their constituents are large and strong.

Condensed matter physics is by far the largest field of contemporary physics. Much progress has also been made in theoretical condensed matter physics. By one estimate, one-third of all US physicists identify themselves as condensed-matter physicists. Historically, condensed-matter physics grew out of solid-state physics, which is now considered one of its main subfields. The term "condensed-matter physics" was apparently coined by Philip Anderson and Volker Heine when they renamed their research group at Cavendish Laboratory — previously "solid-state theory" — in 1967. In 1978, the Division of Solid State Physics at the American Physical Society was renamed as the Division of Condensed-matter Physics. Condensed-matter physics has a large overlap with chemistry, materials science, nanotechnology and engineering.

One of the reasons for calling the field "condensed-matter physics" is that many of the concepts and techniques developed for studying solids actually apply to fluid systems. For instance, the conduction electrons in an electrical conductor form a type of quantum fluid with essentially the same properties as fluids made up of atoms. In fact, the phenomenon of superconductivity, in which the electrons condense into a new fluid phase in which they can flow without dissipation, is very closely analogous to the superfluid phase found in He_4 at low temperatures and He_3 when two of these atoms pair and thus take on boson properties.

Molecular Physics

Molecular physics is the study of the physical properties of molecules and of the chemical bonds between atoms that bind them. Its most important experimental techniques are the various types of spectroscopy. The field is closely related to atomic physics and overlaps greatly with theoretical chemistry, physical chemistry and chemical physics.

Additionally to the electronic excitation states which are known from atoms, molecules are able to rotate and to vibrate. These rotations and vibrations are quantised, there are discrete energy levels. The smallest energy differences exist between different rotational states, therefore pure rotational spectra are in the far infrared region (about 30-150 μm wavelength) of the electromagnetic spectrum. Vibrational spectra are in the near infrared (about 1-5 μm) and spectra resulting from electronic transitions are mostly in the visible and ultraviolet regions.

From measuring rotational and vibrational spectra properties of molecules like the distance between the nuclei can be calculated.

One important aspect of molecular physics is that the essential atomic orbital theory in the field of atomic physics expands to the molecular orbital theory.

Nuclear Physics

Nuclear physics is the field of physics that studies the building blocks and interactions of atomic nuclei.

It must not be confused with atomic physics, that studies the combined system of the nucleus and its arrangement of electrons, even if both terms are sometimes used synonymously in standard English.

Particle physics is a field that has evolved out of nuclear physics and for this reason has been included under the same term in earlier times.

Nuclear power and nuclear bombs are the most commonly known applications of nuclear physics, but the research field is also the basis for a far wider range of less common applications, like, e.g., in the medical sector (nuclear medicine, magnetic resonance imaging), in materials engineering (ion implantation) or archaeology (radiocarbon dating).

Optical Physics

Optical physics, or optical science, is a subfield of atomic, molecular, and optical physics. It is the study of the generation of electromagnetic radiation, the properties of that radiation, and the interaction of that radiation with matter, especially its manipulation and control. It differs from general optics and optical engineering in that it is focused on the discovery and application of new phenomena. There is no strong distinction, however, between optical physics, applied optics, and optical engineering, since the devices of optical engineering and the applications of applied optics are necessary for basic research in optical physics, and that research leads to the development of new devices and applications. Often the same people are involved in both the basic research and the applied technology development.

Researchers in optical physics use and develop light sources that span the electromagnetic spectrum from microwaves to x-rays. The field includes the generation and detection of light, linear and non-linear optical processes, and spectroscopy. Lasers and laser spectroscopy have transformed optical science. Major study in optical physics is also devoted to quantum optics and coherence, and to femtosecond optics.

In optical physics, support is also provided in areas such as the non-linear response of isolated atoms to intense, ultra-short electromagnetic fields, the atom-cavity interaction at high fields, and quantum properties of the electromagnetic field. Other important areas of research include the development of novel optical techniques for nano-optical measurements, diffractive optics, low-coherence interferometry, optical coherence tomography, and near-field microscopy. Research in optical physics places an emphasis on ultrafast optical science and technology. The applications of optical physics create advancements in communications, medicine, manufacturing, and even entertainment.

Particle Physics

Particle physics is a branch of physics that studies the elementary constituents of matter and radiation, and the interactions between them. It is also called *high energy physics*, because many elementary particles do not occur under normal circumstances in nature, but can be created and detected during energetic collisions of other particles, as is done in particle accelerators. Research in this area has produced a long list of particles.

Modern particle physics research is focused on subatomic particles, which have less structure than atoms. These include atomic constituents such as electrons, protons, and neutrons (protons and neutrons are actually composite particles, made up of quarks), particles produced by radiative and scattering processes, such as photons, neutrinos and muons, as well as a wide range of exotic particles.

Strictly speaking, the term *particle* is a misnomer because the dynamics of particle physics are governed by quantum mechanics. As such, they exhibit wave-particle duality, displaying particle-like behaviour under certain experimental conditions and wave-like behaviour in others (more technically they are described by state vectors in a Hilbert space). Following the convention of particle physicists, "elementary particles" refer to objects such as electrons and photons, with the understanding that these "particles" display wave-like properties as well.

All the particles and their interactions observed to date can almost be described entirely by a quantum field theory called the *Standard Model*. The Standard Model has 40 species of elementary particles (24 fermions, 12 vector bosons, and 4 scalar bosons), which can combine to form composite particles, accounting for the hundreds of other species of particles discovered since the 1960s. The Standard Model has been found to agree with almost all the experimental tests conducted to date. However, most particle physicists believe that it is an incomplete description of nature, and that a more fundamental theory awaits discovery. In recent years, measurements of neutrino mass have provided the first experimental deviations from the Standard Model.

Particle physics has had a large impact on the philosophy of science. Some particle physicists adhere to reductionism, a point of view that has been criticised and defended by philosophers and scientists.

Thermodynamics

In physics, thermodynamics is the study of the transformation of energy into different forms and its relation to macroscopic variables such as temperature, pressure, and volume. Its underpinnings, based upon statistical predictions of the collective motion of particles from their microscopic behaviour, is the field of statistical thermodynamics, a branch of statistical mechanics. Roughly, heat means "energy in transit" and dynamics relates to "movement"; thus, in essence thermodynamics studies the movement of energy and how energy instills movement. Historically, thermodynamics developed out of need to increase the efficiency of early steam engines.

The starting point for most thermodynamic considerations are the laws of thermodynamics, which postulate that energy can be exchanged between physical systems as heat or work. They also postulate the existence of a quantity named entropy, which can be defined for any system. In thermodynamics, interactions between large ensembles of objects are studied and categorised. Central to this are the concepts of system and surroundings. A system is composed of particles, whose average motions define its properties, which in turn are related to one another through equations of state. Properties can be combined to express internal energy and thermodynamic potentials, which are useful for determining conditions for equilibrium and spontaneous processes.

With these tools, thermodynamics describes how systems respond to changes in their surroundings. This can be applied to a wide variety of topics in science and engineering, such as engines, phase transitions, chemical reactions, transport phenomena, and even black holes. The results of thermodynamics are essential for other fields of physics and for chemistry, chemical engineering, aerospace engineering, mechanical engineering, cell biology, biomedical engineering, materials science, and economics to name a few.

Fundamental Theories

Gaseous State

The molecules of a gas are very much apart from one another. Hence, they exert practically no attraction on one another. Therefore, they are completely free to move over the entire space open to them and fill the vessel completely in which the gas is contained. This is why a gas has neither a definite size nor a definite shape.

Gas Pressure: The pressure exerted by a gas on the walls of the containing vessel is explained as due to the molecules constantly colliding with the wall and rebounding from it. The rate of change of momentum thus produced over unit area of the wall gives a measure of the pressure exerted by the gas.

If, at a given temperature, the volume of the vessel is doubled, the number of collisions at its walls in a given time will be halved, *i.e.* its pressure will be halved. This is Boyle's law. Further, if the temperature of the gas is raised at constant volume, the molecules will gain velocity.

Then the number of collisions at the walls per second, and also the change in momentum in each collision, will increase. It means that the pressure of the gas will increase.

Temperature-rise on Compression: When a gas is compressed, the work done is added to the kinetic energy of its molecules. Hence, the temperature of the gas rises.

Gases vs Kinetic Theory

According to the kinetic theory of gases:

(i) A gas is made up of a very large number of identical small particles, called molecules.

(ii) These molecules are constantly moving in straight lines in all directions with various high speeds.

(iii) They are colliding with one another, and with the walls of the containing vessel so that the speed and direction of any particular molecule are frequently changing. As a result of these collisions the average density of molecules is the same everywhere in the container.

(iv) The gas exerts pressure on the walls of the containing vessel. This pressure arises due to the collisions of the molecules with the walls of the vessel. As a molecule collides, it suffers a change in momentum. The rate of change of momentum is equal to the force exerted on the wall (Newton's second law of motion). Since a large number of molecules is colliding frequently, they exert a steady force given by the average rate of change of momentum. This force measured per unit area of the wall is called the "pressure" of the gas.

Basic Assumptions of Kinetic Theory and Pressure of an Ideal Gas: The kinetic theory leads to an expression for the pressure of a gas. For this, we make the following assumptions:

(i) *The Molecules of a Gas are Hard, Smooth and Spherical Particles:* Hence, their rotation is insignificant.

(ii) *They are Extremely Small:* Hence, the volume of the molecules themselves is a negligibly small fraction of the volume occupied by the gas.

(iii) *They Exert no Appreciable Force on One Another Except during a Collision:* Thus, between collisions, the molecules move with constant speeds in straight lines.

(iv) *All the Collisions are Perfectly Elastic:* Thus, there is no loss of kinetic energy during a collision, and a molecule rebounds from the wall with a speed equal to that with which it strikes.

(v) *The Duration of a Collision is Negligible Compared with the Time Spent by a Molecule between Collisions:* Thus, the kinetic energy which is converted into potential energy of deformation during the collision is available again as kinetic energy after such a brief time that we can ignore the exchange entirely.

Let us consider an ideal gas contained in a cubical vessel with perfect elastic walls. Let l be the length of each edge. Let m be the mass of a molecule and n the total number of molecules in the gas. These molecules are moving in all directions with different speeds. These speeds are distributed about an average speed which depends on the temperature of the gas. This is known as "Maxwellian distribution of speeds."

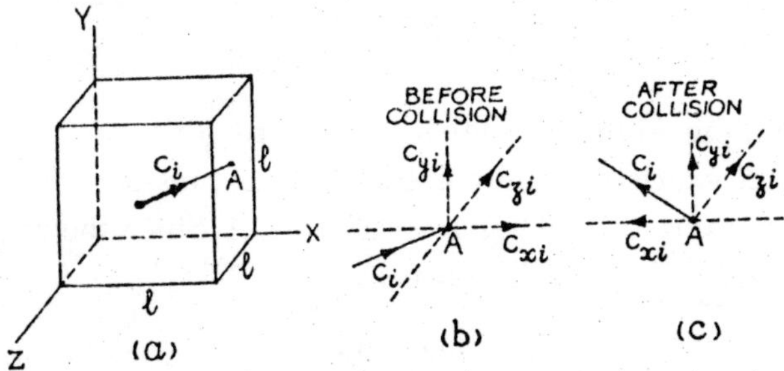

Let us consider a particular molecule i approaching the face A of the cube with a velocity $\vec{c_i}$. We can resolve $\vec{c_i}$ into three velocity components c_{xi}, c_{yi}, c_{zi} along the x, y and z-axis parallel to three adjacent edges of the cube, such that

$$C_i^2 = c_{xi}^2 + c_{yi}^2 + c_{zi}^2. \qquad ...(i)$$

If this particle collides with the face A which is perpendicular to x-axis, only the component velocity c_{xi} will be affected. Since the collision is perfectly elastic, the molecule will rebound with the component c_{xi} reversed in direction as shown in Fig. Thus, its momentum along x-axis changes from $m\ c_{xi}$, to — $m\ c_{xi}$, *i.e.*, the change in the momentum of this molecule due to the collision is

$$-m\ c_{xi} - (m\ c_{xi}) = -\ 2m\ c_{xi}.$$

Hence, the momentum imparted to A will be $2m\ c_{xi}$, as the total momentum is conserved. The molecule now strikes the face opposite to A, rebounds from it and again strikes A. The time between two successive collisions on A is $2l/c_{xi}$, since $2l$ is the distance travelled along the x-axis with a velocity c_{xi}. The number of collisions per second is, therefore, $c_{xi}/2l$. Hence, the rate at which the molecule transfers momentum to A is

$$2m\ c_{xi} \times \frac{c_{xi}}{2l} = \frac{mc_{xi}^2}{l}.$$

Since there are n molecules, the rate at which momentum is imparted to the face A by all the molecules is

$$\sum_{i=1}^{n} \frac{mc_{xi}^2}{l}$$

According to the Newton's second law, this rate is equal to the force exerted. Therefore, the above expression gives the total force exerted by the gas on the face A. Since the area of the face is l^2, the pressure on it is

$$px = \frac{m}{l^3} \sum_{i=1}^{n} c_{xi}^2.$$

Similarly, the pressures exerted on the other two faces are

$$p_y = \frac{m}{l^3}\sum_{i=1}^{n} c_{yi}^{\ 2} \quad \text{and} \quad p_z = \frac{m}{l^3}\sum_{i=1}^{n} c_{zi}^{\ 2}$$

But the pressure on all the faces of the cube is same, say p. Thus, we can write

$$p = p_x = p_y = p_z$$

or

$$p = \frac{1}{3}\ (p_x + py + pz)$$

$$= \frac{1}{3}\ \frac{m}{l^3}\left[\sum_{i=1}^{n} c_{xi}^{\ 2} + \sum_{i=1}^{n} c_{yi}^{\ 2} + \sum_{i=1}^{n} c_{zi}^{\ 2}\right]$$

$$= \frac{1}{3}\ \frac{m}{l^3}\sum_{i-1}^{n}\left(c_{xi}^{\ 2} + c_{yi}^{\ 2} + C_{zi}^{\ 2}\right)$$

$$= \frac{1}{3}\ \frac{m}{l^3}\sum_{i=1}^{n} c_{ci}^{\ 2} \qquad \text{[by eq. (i)]}$$

$$= \frac{1}{3}\ \frac{m}{V}\ \sum_{i=1}^{n} c_i^{\ 2}$$

where V (= l^3) is the volume of the cube.

Now, if $\overline{c^2}$ be the *mean of the squares of the speed of all the molecules,* we have

$$\overline{c^2} = \frac{\sum_{i=1}^{n} c_i^{\ 2}}{n}$$

$$\sum_{i=1}^{n} c_i^{\ 2} = n\overline{c^2}$$

$\overline{c^2}$ is called the 'mean-square-speed' of the molecules. Hence, the last expression gives

$$\boxed{p = \frac{1}{3}\frac{mn}{V}\overline{c^2}} \qquad \text{...(ii)}$$

Now $\frac{mn}{V} = \frac{\text{mass of the gas}}{\text{volume of the gas}}$ = density ρ say

$$\boxed{p = \frac{1}{3}\rho\overline{c^2}.} \qquad \text{...(iii)}$$

This may be put as

$$p = \frac{2}{3}\left(\frac{1}{2}\rho\,\overline{c^2}\right).$$

But $\frac{1}{2}\rho\,\overline{c^2}$ is the kinetic energy per unit volume of the gas. Hence, *the pressure is two-third of the total translational kinetic energy of the molecules per unit volume.*

This means that if the gas is compressed, the kinetic energy of translation of its molecules will increase which is proportional to the absolute temperature. Hence, the temperature of the gas will rise.

Role of the 'Large' number of Molecules of a Gas: According to kinetic theory, a gas consists of a *large* number of molecules moving in all directions with various speeds which are constantly changing due to the frequent collisions with one another and with the walls of the containing vessel. It is impossible to study and apply the laws of mechanics to the motion of individual molecules. We can, however, apply the laws of mechanics to the average behaviour of molecules. For example, we calculate the pressure of the gas by calculating the 'average' rate of transfer of momentum to the walls by the colliding molecules. This average is well-defined only because the number of molecules is very large and hence collisions are taking place constantly. Hence, it is the 'large' number of molecules which helps us in expressing thermodynamic properties of a microscopic system like a gas by applying the laws of mechanics statistically to the system.

Gravitational Potential Energy of Molecules: The molecules of a gas are in continuous rapid motion with respect to the centre of mass of the gas, but the number of molecules is so 'large' that their spatial distribution in the gas may be treated to remain unchanged. Hence, their gravitational potential energy may also be treated to remain unchanged. Furthermore it is very much smaller than kinetic energy.

Force Exerted by the Molecules: Again, it is due to their 'large' number that the molecules exert a 'steady' force on the walls of the containing vessel by colliding with them. As a molecule collides, it suffers a change in momentum. The rate of change of momentum is equal to the force exerted on the wall (Newton's second law of motion).

Since a large number of molecules are colliding frequently, they exert a *steady* force given by the average rate of change of momentum.

Dependence of Pressure on the 'square' of the Molecular Speed The pressure exerted by a gas on the walls of the containing vessel is the average rate of transfer of momentum to the unit area of the walls by the colliding molecules. Now, both the momentum transferred in one collision and the number of collisions per second depend on the molecular speed. Hence, the pressure depends upon the 'square' of the molecular speed.

Approach of Real Gases towards Ideal Gas at Low Pressure: Experiments show that real gases approach an ideal gas as the pressure (density), *i.e.* the number of molecules per unit volume is decreased. This does not contradict the kinetic theory which assumes a large number of molecules in a gas. Actually, a gas consists of $\frac{6\times10^{23}}{22.4}$ molecules per litre at N. T. P. so that when the pressure is sufficiently decreased, the number of molecules per litre is still very large. Moreover, at low pressure the molecules become further apart so that the force of attraction between them decreases. This is a support to the kinetic theory which assumes no force of attraction between the molecules except during a collision.

Inelastic Collisions: The inelastic collisions of molecules of a substance, say a gas, with each other do not involve a loss of mechanical energy to heat. The reason is that the mechanical energy of the molecules itself is heat. Therefore, any loss in the mechanical energy of the molecules in an inelastic collision means a corresponding gain in heat which is nothing but the mechanical energy itself.

The mechanical (kinetic) energy of a moving bullet is, however, not the heat in the bullet. (It is the mechanical energy associated with the *kinetic motion of the molecules* of the bullet with respect to its centre of mass, which is the heat in the bullet). Hence, when the bullet strikes a target, its mechanical energy is lost as heat so that its temperature rises (the kinetic motion of its molecules increases).

Inelastic Walls: If the gas molecules make inelastic collisions with the walls of the containing vessel, some kinetic energy may be lost as heat. But if the walls are at the same temperature as the gas, this heat will come back to the molecules as kinetic energy. Hence, inelastic collisions do not matter so long as the walls are at the same temperature as the gas.

Root Mean's Square Speed

According to the kinetic theory, a gas is made up of a very large number of small particles, called molecules. These molecules are constantly moving in straight lines in all directions. All the molecules, however, do not move with the same speed, but the speeds are distributed about an average value depending upon the temperature of the gas.

Although each molecule is having frequent collisions in which its speed is changed but, Maxwell has shown, the number of molecules having speeds between any given range remains unchanged.

If $c_1, c_2, c_3, \ldots c_n$ be the speeds of n molecules of the gas, then the mean of the squares of these speeds is called the mean-square-speed $\overline{c^2}$ of the molecules. Thus,

$$\overline{c^2} = \frac{c_1^{\,2} + c_2^{\,2} + \ldots + c_n^{\,2}}{n}$$

The square-root of $\overline{c^2}$ *i.e.* $\sqrt{\overline{c^2}}$ is called the 'root-mean-square (rms) speed' of the molecules and may be indicated as c_{rms}. From kinetic theory, the pressure exerted by an ideal gas is given by

$$p = \frac{1}{3}\rho\overline{c^2}$$

or

$$\sqrt{\overline{c^2}} = c_{rms} = \sqrt{\frac{3p}{\rho}} \qquad \ldots(1)$$

We know that the speed of sound in a gas is given by

$$V = \sqrt{\frac{\gamma p}{\rho}} \text{ (Laplace's formula).}$$

Comparing this with eq. (i), we see that the speed of sound in a gas, v, is of the same order as the *rms* speed of the gas molecules, c_{rms}. In fact

$$\text{v} \propto c_{rms}$$

Thus, if v_1 and v_2 be the sound speeds in two gases whose molecules have *rms* speeds c_{1rms} and c_{2rms} then

$$\frac{v_1}{v_2} = \frac{c_{1rms}}{c_{2rms}}$$

Relation between rms Speed and Molecular Weight: The density of 1 mole of a gas is

$$p = \text{M/V}, \qquad \ldots(\text{ii})$$

where M is the molecular weight of the gas (*i.e.* the mass of 1 mole of the gas in gram) and V the volume. Further, the ideal gas equation for 1 mole is

$$pV = RT. \qquad \text{...(iii)}$$

Dividing eq. (iii) by eq. (ii), we get

$$\frac{p}{\rho} = \frac{RM}{M}$$

Hence, from eq. (i), we obtain

$$\boxed{c_{rms} = \sqrt{\frac{3RT}{M}}.}$$

Thus, *the rms speed of the molecules of an ideal gas is inversely proportional to the square-root of the molecular weight of the gas.*

Now, $M = mN$, where m is the mass of a single molecule and N (Avogadro's number) is the number of molecules per mole of a gas. Then, we have from the last expression,

$$c_{rms} = \sqrt{\frac{3RT}{mN}} = \sqrt{\frac{3kT}{m}},$$

where k (= R/N) is Boltzmann's constant.

At a given temperature T, this means that

$$c_{rms} \propto \sqrt{\frac{1}{m}},$$

because k is (universal) constant. Thus, if we have two gases whose *rms* speeds are c_{1rms} and C_{2rms} and whose molecules are of masses m_1 and m_2, then

$$\frac{c_{1rms}}{c_{2rms}} = \sqrt{\frac{m_2}{m_1}}$$

That is, at the same temperature, *the ratio of the root-mean-square speeds of molecules of two different gases is equal to the square-root of the inverse ratio of their masses.*

Application in Separation of Isotopes: The above relation leads to important fact regarding the diffusion of two different gases through the porous walls of a container. The lighter gas whose molecules have a larger speed, would escape faster than the heavier gas. The diffusion process can, therefore, be used to separate the lighter isotope from a normal specimen of a material.

We know that ordinary uranium consists of a lighter isotope U^{235} (0.7%) and a heavier isotope U^{238} (99.3%). The lighter isotope U^{235} is very useful in making atom bomb because its nucleus can be comparatively much easily fissioned. Hence, U^{235} must be separated from ordinary uranium. This can be done by allowing a gaseous compound of uranium to diffuse through the porous walls of a container. The part of the gas diffused through the walls in a small time-interval will be richer in U^{235}. The process may be repeated by using a number of successive diffusion stages.

Air Leaking into Vacuum is Richer in H_2 and He: H_2 and He are much lighter than other gases N_2, O_2, etc. found in air, and so have larger root-mean-square speeds. Hence, they leak faster.

Time for Diffusion-Separation: In order to separate two gases from a mixture by diffusion, a number of successive diffusion stages are used, the diffusion time in each stage being *small.* If time would be large, both gases will ultimately diffuse out and separation would not be possible.

Kinetic Interpretation of Temperature: Maxwell worked out the dynamical theory of gas and arrived at an important result according to which the average kinetic energy of translation of the molecules of an ideal gas is directly proportional to its absolute temperature. This result is known as the "kinetic interpretation of temperature".

Let us consider 1 gram-molecule of a gas at absolute temperature T and having volume V. The number of molecules in one mole is N (Avogadro's number). If m is the mass of a molecule and $\overline{c^2}$ the mean-square speed, then its kinetic pressure is given by

$$p = \frac{1}{3}\rho\overline{c^2} = \frac{1}{3}\frac{mN}{V}\overline{c^2}$$

or $$pV = \frac{2}{3} N\left(\frac{1}{2}m\overline{c^2}\right). \quad \text{...(i)}$$

But $\frac{1}{2}m\overline{c^2}$ is the average kinetic energy of translation of a molecule which, according to the kinetic interpretation of temperature, is directly proportional to T. Thus,

$$\frac{1}{2}m\overline{c^2} = \alpha T, \quad \text{...(ii)}$$

where α is a constant. Therefore, from eq. (i), we get

$$pV = \frac{2}{3} N\alpha T.$$

Comparing it with the gas equation $pV= RT$, we find that

$$\frac{2}{3}N\alpha = R \text{ or } \alpha = \frac{3}{2}\frac{R}{N}.$$

The ratio R/N is (universal) *gas constant per molecule* and is called "Boltzmann's constant" denoted by k. Its unit is joule/K.

$$\therefore \quad \alpha = \frac{3}{2}k.$$

Putting this value of α in eq. (ii), we get

$$\boxed{\frac{1}{2}m\overline{c^2} = \frac{2}{3}kT.}$$

Thus, *the average kinetic energy of translation per molecule of an ideal gas is* $\frac{2}{3}$ *k T, whatever may be the mass of the individual molecules.* If we put 7= 0 in the above expression, then

$$\overline{c^2} = 0.$$

Hence, *according to the kinetic interpretation, the absolute zero of temperature is the temperature at which the molecular motion ceases*. This is, however, not strictly correct because at absolute zero the molecules do have some energy, called 'zero-point energy'.

Inner Energy and Measurement: The temperature of a body is related to the total translational kinetic energy of the molecules, and not to the potential energy. The internal energy of the body, however, includes all forms of molecular energies.

Cold in the Upper Atmosphere: The gas kinetic temperature in the upper atmosphere is of the order of 1000 K and as such the average kinetic energy of translation *per molecule* ($\frac{3}{2}kT$) is higher as compared to that on the earth. But the density of the upper atmosphere is much lower, so that the number of molecules per unit volume is much smaller than on the earth. Therefore, the total kinetic energy of translation of the molecules per unit volume, and hence the total heat content per unit volume, is smaller. Hence, we feel quite cold there.

Important Gas Laws from Kinetic Theory

We can deduce gas laws from the kinetic theory.

Boyle's Law: According to this law, the product of the pressure and volume of a given mass of gas at constant temperature is constant. The kinetic theory expression for the pressure of a given mass of gas is (in usual notations)

$$p = \frac{1}{3}\frac{mn}{V}\overline{c^2} \quad \text{or} \quad pV = \frac{1}{3}mn\overline{c^2}$$

If the temperature is constant, the average kinetic energy of translation per molecule is constant. Hence, the mean-square-speed c^2 is constant. Also, the mass of gas *mn*, is constant. Hence, the last expression gives pV = constant,

This is Boyle's law.

Charle's Law: According to this law, the volume of a given mass of gas at constant pressure is directly proportional to the absolute temperature. From kinetic theory, the pressure of a given mass of the gas is

$$p = \frac{1}{3}\frac{mn}{V}\overline{c^2} \quad \text{or} \quad V = \frac{2}{3}\frac{n}{p}\frac{1}{2}m\overline{c^2}$$

But the mean kinetic energy of a molecule $\frac{1}{2}m\overline{c^2} = \frac{3}{2}kT.$

$$\therefore \qquad V = \frac{2}{3}\frac{n}{p}\frac{3}{2}kT = \frac{n}{p}kT$$

If p be constant, then

$$V \propto T.$$

This is Charle's law.

(iii) *Avogadro's Law:* According to this law, equal volumes of all gases under the same conditions of temperature and pressure contain equal numbers of molecules.

Let two equal volumes V of different gases contain, respectively n_1 and n_2 molecules of masses m_1 and m_2. If p be the pressure exerted by each gas, then by the kinetic theory, we have

$$pV = \frac{1}{3}m_1 n_1 \overline{c^2}$$

and

$$pV = \frac{1}{3}m_2 n_2 \overline{c^2},$$

where $\overline{c_1^2}$ and $\overline{c_2^2}$ are, respectively the mean-square-speeds of the molecules of the two gases. These equations show that

$$m_1\, n_1\, \overline{c^2} = m_2\, n_2\, \overline{c_2^2}$$

Now, if the gases are at the same temperature, their average kinetic energies of translation per molecule are equal. Thus, $\frac{1}{2}m_1, \overline{c_2^2} = \frac{1}{2}m_2\overline{c_2^2}$ Applying this result in the last expression, we get $n_1 = n_2$ which establishes Avogadro's law.

(iv) *Dalton's Law of Partial Pressures:* According to this law, the total pressure exerted by a mixture of non-reacting gases occupying a vessel is equal to the sum of the individual pressures which each gas would exert if it alone occupied the whole vessel.

Let us consider the individual gases in a mixture which occupies a volume V. Suppose the first gas contains n_1 molecules, each of mass m_1 and mean-square-speed $\overline{c_1^2}$, the second gas contains n_2 molecules each of mass m_2 and mean-square-speed $\overline{c_2^2}$ and so on. Let p_1, p_2, ...respectively be the partial pressures of the gases. Each gas fills the whole volume V at its partial pressure. Thus, according to the kinetic theory, we have

$$p_1 V = \frac{1}{3} m_1 n_1 \overline{c^2},$$

$$p_2 V = \frac{1}{3} m_2 n_2 \overline{c^2} \quad \text{and so on.}$$

Adding, we get

$$(P_1 + P_2 + \ldots\ldots) V = \frac{1}{3} (m_1 n_1 \overline{c_1^2} + m_2 n_2 \overline{c_2^2} \ldots\ldots) \quad \text{(i)}$$

Now, the entire mixture is at a uniform temperature. Therefore, the average kinetic energy of translation per molecule of each gas is the same.

This means that

$$\frac{1}{2} m_1 \overline{c_1^2} = \frac{1}{2} m_2 \overline{c_2^2} \ldots\ldots \frac{1}{2} m \overline{c^2} \text{ .(say)}$$

Substituting this result in eq. (i), we have

$$(P_1 + P_2 + \ldots\ldots) V = \frac{1}{3} (n_1 + n_2 + \ldots\ldots) m \overline{c^2} \quad \ldots\text{(ii)}$$

Now, the mixture has a total number of molecules $(n_1 + n_2 + ...)$. Hence, the pressure p exerted by the mixture must be given by

$$pV = \frac{1}{3}(n_1 + n_2 + ...)\, m\overline{c^2}. \qquad ...(iii)$$

Eq. (ii) and (iii) show that

$$P = P_1 + P2 + ...$$

This is Dalton's law of partial pressures.

Independence of Sound Speed from Pressure and Density : The speed of sound in a gas is given by

$$V = \sqrt{\frac{\gamma p}{\rho}}, \qquad ...(iv)$$

where $\overline{c^2}$ is the pressure and ρ the density of the gas. From kinetic theory, the pressure of the gas is given by

$$p = \frac{1}{3}\rho\,\overline{c^2}$$

or

$$\frac{p}{\rho} = \frac{1}{3}\overline{c^2} \qquad ...(v)$$

where $\overline{c^2}$ is the mean-square-speed of gas molecules. According to the kinetic interpretation of temperature, so long the temperature remains constant, $\overline{c^2}$ remains constant, therefore p/ρ also remains constant. Hence, *the speed of sound in a gas is independent of pressure and density of the gas at constant temperature*

Squaring eq. (iv) and putting the value of p/ρ from eq. (v), we get

$$v^2 = \gamma\left(\tfrac{1}{3}\overline{c^2}\right)$$

or

$$\frac{v^2}{\overline{c^2}} = \gamma/3.$$

This is what we had to show.

Degrees of Freedom

The degrees of freedom of a particle indicate the number of *independent* motions which the particle can undergo, or the number of *independent* methods of exchanging energy.

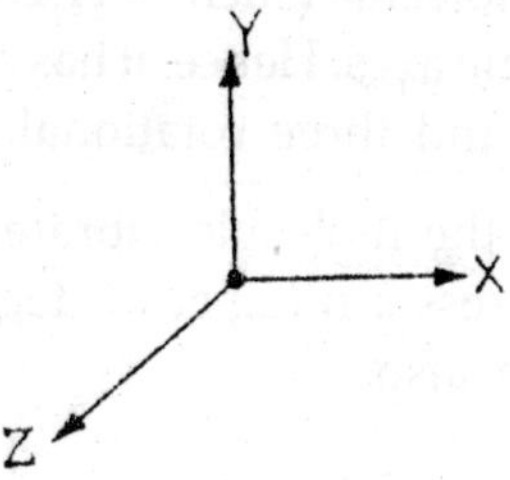

The molecule of a monatomic gas (such as He, A, etc.) consists of a single atom. Its translational motion can take place in any direction in space. Thus, it can be resolved along three coordinate axes and can have three independent motions. Hence, it has *three* degrees of freedom, all translational. (A monatomic molecule can rotate also. But it has such a small moment of inertia that its kinetic energy of rotation is insignificant. Therefore, it possesses only kinetic energy of translation).

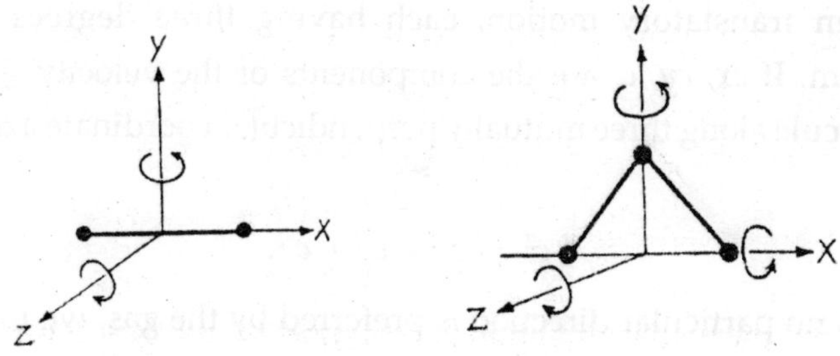

The molecule of a diatomic gas (such as H_2, O_2, etc.) is made up of two atoms joined rigidly to one another through a bond. This can not only move bodily, but also rotate about any one of the three coordinate axes. However, its moment of inertia about the axis joining the two atoms is negligible compared to that about the other two axes. Hence, it can have only two rotational motions. Thus, a diatomic molecule has five degrees of freedom, three with respect to translation and two with respect to rotation.

A polyatomic molecule (such as H_2O) can rotate about any of the three coordinate axes. Hence, it has *six* degrees of freedom, three translational and three rotational.

If the atoms of the molecule vibrate with respect to each other, it would possess a number of degrees of freedom with respect to vibration also.

Equal Partition vs Energy

According to this law, the total internal energy of a classical dynamical system consisting of a large number of particles (as a gas) is distributed equally among its various degrees of freedom, and for a system in equilibrium at temperature T, the average energy per particle (having quadratic term) associated with each degree of freedom is-kT, where k is Boltzmann's constant.

Proof: Let us consider one mole of a gas at absolute temperature T and having volume V. Its molecules are in random translatory motion, each having three degrees of freedom. If cx, cy, c_z are the components of the velocity $\vec{c}$ of a molecule along three mutually perpendicular coordinate axes, then

$$c^2 = c_x^2 + c_y^2 + c_z^2.$$

As no particular direction is preferred by the gas, we may write

$$\overline{c_x^2} = \overline{c_y^2} = \overline{c_z^2} = \frac{1}{3}\,\overline{c^2}, \qquad \text{...(i)}$$

where $\overline{c_x^2}$, $\overline{c_y^2}$ and $\overline{c_z^2}$ are the average values of the squares of the components (along the three axes) of all the molecules of the gas, and $\overline{c^2}$ is average value of c^2 for all the molecules.

Now, from kinetic theory, we have

$$pV = \frac{1}{3}\ \mathrm{mN}\overline{c^2},$$

where N is total number of molecules in one mole, *i.e.*, Avogadro's number. Using gas law *($pV = RT$)*, we may write

$$\frac{1}{3}\mathrm{mN}\overline{c^2} = RT$$

or $$\frac{1}{2}\mathrm{m}\overline{c^2} = \frac{3}{2}\frac{R}{N}T = \frac{3}{2}kT, \qquad \text{...(ii)}$$

where k (= R/N) is Boltzmann's constant.

Now, according to eq. (i), the average kinetic energy of a gas molecule in each translational degree of freedom is

$$\frac{1}{2}\overline{c_x^{\ 2}}\ \frac{1}{2}\overline{c_y^{\ 2}}\ \frac{1}{2}m = m = m\overline{c_z^{\ 2}} = \frac{1}{3}\left(\frac{1}{2}m\overline{c^2}\right)$$

But, by eq. (ii), $\frac{1}{2}m\overline{c^2} = \frac{3}{2}kT.$

$$\therefore \qquad \frac{1}{2} = m\overline{c_x^{\ 2}} = \frac{1}{2}\overline{c_y^{\ 2}}\ \frac{1}{2}\overline{c_z^{\ 2}}\ \frac{1}{2}m = m = kT.$$

Thus, the average kinetic energy of a gas molecule per degree of freedom is $\frac{1}{2}$ kT. Maxwell arrived at this result for translatory motion. Later on, Boltzmann showed that the average kinetic energy of a molecule in each rotational degree of freedom, as well as the average kinetic energy, and also the average potential energy (having quadratic term), in each vibrational degree of freedom is also $\frac{1}{2}$ kT.

This is the law of equipartition of energy.

Degrees of Freedom

The degrees of freedom of a particle is the number of 'independent' motions which the particle can undergo, or the number of ' independent ' methods of exchanging energy. A monatomic molecule has three degrees of freedom (all translational), a diatomic one has five (three translational plus two rotational) while a triatomic one has six (three translational plus three rotational).

Law of Equipartition of Energy: According to this law, in a classical system of particles, which is in equilibrium at absolute temperature T, the average internal energy per particle (having a quadratic term) associated with each degree of freedom is $\frac{1}{2}kT$, where k is Boltzmann's constant. If the particle has f degrees of freedom, its average energy would be $\frac{1}{2}fkT$.

Molar Specific Heats and the Ratio of Specific Heats of Gases: Let us consider 1 mole of an ideal gas at Kelvin temperature T. It has N molecules, where N is Avogadro's number. Since in an ideal gas there are no forces of attraction between the molecules, there is no internal potential energy. The internal energy of an ideal gas is entirely the kinetic energy (having quadratic terms) of its molecules. Now, the average kinetic energy per molecule of an ideal gas is

$\frac{1}{2}fkT$, where f is the number of degrees of freedom. Therefore, the internal energy of one mole of an ideal gas would be

$$U = N \times \frac{1}{2}fkT = \frac{1}{2}fRT. \qquad [\because k = R/N]$$

Differentiating it, we get

$$\frac{dU}{dT} = \frac{1}{2}fR.$$

Suppose the gas is heated at constant volume until its temperature rises by dT. The heat supplied would be $dQ = C_V dT$, and the external work would be zero (as volume remains constant). Therefore, by the first law of thermodynamics, $dU = dQ\text{-}dW$, we have

$$dU = C_V dT$$

$$C_V = \frac{dU}{dT}$$

Substituting the value of dU/dT from above, we get

$$\boxed{C_V = \frac{f}{2}R.}$$

This is the molar specific heat of the gas at constant volume. The molar specific heat at constant pressure would be

$$C_p = C_V + R \qquad [\therefore C_p\text{-}C_V = R]$$

$$= \frac{f}{2}R + R$$

or

$$\boxed{C_p = \left(\frac{f}{2}+1\right)R.}$$

Hence, the ratio of the two specific heats of a gas would be

$$\gamma \frac{C_p}{C_V} = \frac{\left(\frac{f}{2}+1\right)R.}{\frac{f}{2}R}$$

or

$$\boxed{\gamma = 1+\frac{2}{f}.}$$

Thus, γ decreases with increase in degrees of freedom.

Let us now predict the values of C_V, C_p and γ for monatomic, diatomic and triatomic gases separately. We shall ignore the vibrational energy which is generally small for gases at ordinary temperatures.

Monatomic Gases: A molecule of monatomic gas has only 3 (translational) degrees of freedom, *i.e.* $f = 3$. Therefore

$$C_V = \frac{f}{2}R = \frac{3}{2} \simeq 3\text{cal/mole-K.}$$

$$(\because R \simeq 2 \text{ cal/mole-K})$$

$$C_p = \left(\frac{f}{2}+1\right)R = \frac{5}{2}R \simeq 5 \textit{ cal/mole-K}$$

$$\gamma = 1+\frac{2}{f} = 1+\frac{2}{3} = \frac{5}{3} = 1.66.$$

The experimental values of C_V, C_p and γ for all the inert and other monatomic gases are in close agreement with the above theoretically deduced values.

Importance of Diatomic Gases: A molecule of diatomic gas has 5 degrees of freedom, 3 translational and 2 rotational, *i.e.* $f = 5$. Therefore

$$C_v = \frac{f}{2} R = \frac{5}{2} R \simeq 5\text{cal/mole-K.}$$

$$C_p = \left(\frac{f}{2}+1\right)R = \frac{7}{2}R \simeq 7\text{cal/mole-K.}$$

$$\gamma = 1+\frac{2}{f} = 1+\frac{2}{5} = \frac{7}{5} = 1.40.$$

The experimental values of Cv, C_p and γ for many diatomic gases like H_2, O_2, N_2 are in close agreement with these theoretical values, but in some cases there is a disagreement.

For example, the experimental values of C_v and C_p for Cl_2 are higher and that of γ lower than the theoretical values.

Importance of Triatomic Gases: A molecule of triatomic gas has 6 degrees of freedom, 3 translational and 3 rotational, *i.e.*, $f = 6$. Therefore

$$C_V = \frac{f}{2} R = 3R \simeq 6 \text{ cal/mole-K.}$$

$$C_p = \left(\frac{f}{2}+1\right) R = 4R \simeq 8 \text{ cal/mole-K.}$$

$$\gamma = 1 + \frac{2}{f} = 1 + \frac{2}{6} = \frac{4}{3} = 1.33.$$

In case of triatomic gases sufficient disagreement is found between the experimental and the theoretical values of C_V, C_p and γ. The experimental specific heats are larger than the theoretical values, whereas γ is smaller.

Thus, the kinetic theory could account for the specific heats of many gases but not for all. Some improvement in the theory can, however, be made by including the vibrational energy of the atoms in diatomic and triatomic molecules. Further, the theory is unable to explain the variation of specific heats with temperature which is actually observed. Hence, the kinetic theory of specific heats has been replaced by the quantum theory of specific heats which explains the observed facts satisfactorily.

Observed Value of C_v for SO_2 gas is 7.5 cal/(mole-k) at Ordinary Temperature and Pressure

SO_2 is a triatomic gas, its molecules having 6 degrees of freedom, 3 translational and 3 rotational. Therefore, according to kinetic theory, its molar specific heat at constant volume, C_v, must be *3R, i.e.* 6 cal/mole-K. The experimental value is 7.5 cal/mole-K which is well above the theoretical value. The excess can be accounted for by supposing that in such molecules the atoms are capable of vibration at ordinary temperature. The complete explanation can, however, be given by the quantum theory of specific heats.

Equal Masses of a Monatomic and a Diatomic Gas: The molecules of a monatomic gas possess only translational degrees of freedom, while those of diatomic gas possess rotational also (and may possess vibrational also). Therefore, the specific heat of monatomic gas is smaller. Hence, when supplied equal quantities of heat, the monatomic gas undergoes a larger temperature-rise.

Variation of Molar Specific Heat, C_v, of Hydrogen with Temperature T: According to kinetic theory, the specific heats of gases, C_V and C_p, must be independent of temperature. In practice, however, the specific heats do vary with temperature. A graph showing the variation of C_V of hydrogen with temperature is drawn in Fig. It shows that for hydrogen (a diatomic gas) C_V agrees with the theoretical value $5R/2$ only in the temperature-range from 250 K to 750 K. Below 250 K the experimental value of C_V decreases to $3R/2$, while above 750 K it increases to $7R/2$. A possible explanation for this variation can be given as follows:

(i) At low temperatures, say upto 60 K, the hydrogen molecule has only motion of translation, *i.e.* hydrogen behaves simply like a monatomic gas. Therefore, the molecule has only 3 degrees of freedom, and so by the law of equipartition of energy, its total energy is $\frac{3}{2}kT$. Hence, the internal energy of 1 mole of gas is

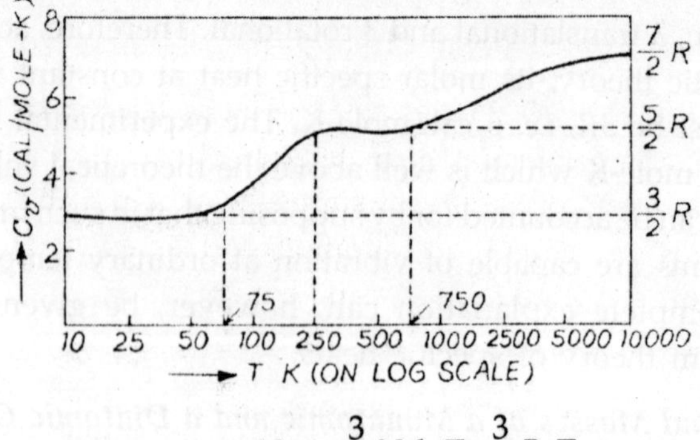

$$U = \frac{3}{2} N k T = \frac{3}{2} R T.$$

The molar specific heat at constant volume is therefore

$$C_V = \frac{dU}{dT} = \frac{3}{2}R,$$

as observed experimentally.

(ii) Above 60 K, the molecule begins to rotate also behaving like a dumbbell, and at about 250 K the rotational energy attains its full classical value. Now, the molecule has 5 degrees of freedom (3 translational + 2 rotational) and so its total internal energy is $\frac{5}{2}\ kT$. Hence, the internal energy of 1 mole of gas is

$$U = \frac{5}{2}NkT = \frac{5}{2}RT.$$

$$\therefore \qquad C_V = \frac{dU}{dT} = \frac{5}{2}R.$$

We observe that between 250 K and 750 K the molar specific heat is $\frac{5}{2}\ R$, as predicted theoretically.

(iii) Above 750 K, the molecular collisions become so vigorous that the atoms in the molecule begin to vibrate so that C_v begins to increase. The vibration is fully excited at about 4000 K when the vibrational energy attains its full value $k\ T$. Thus, the total energy (translational + rotational + vibrational) of the molecule becomes $\frac{3}{2}\ \frac{7}{2}kT + kT + kT = kT$, so that the total internal energy of 1 mole of gas is

$$U = \frac{7}{2}NkT = \frac{7}{2}RT.$$

$$\therefore \qquad C_V = \frac{dU}{dT} = \frac{7}{2}$$

We observe that above 4000 K the molar specific heat is $\frac{7}{2}R$, as predicted theoretically.

Different gases show such change-overs at different temperatures because their molecular structures are different. For example, in a Cl_2 molecule atomic vibrations are excited even at the room temperature.

Although we have given an explanation for the variation of specific heat with temperature but it contradicts the kinetic theory result that the specific heat of gases must be independent of temperature. Only a complete quantum discussion of H_2 molecule can explain its above specific heat behaviour satisfactory.

PROBLEMS

1. The number of molecules per cu cm of a gas is 2.7×10^{19} at 0°C and 76 cm of mercury. What is the number per cm at (i) 0°C and 10^{-6} mm of mercury and (ii) 39°C and 10^{-6} mm of mercury ?

Solution: By kinetic theory, the pressure per unit volume of a gas is given by $p = \frac{1}{3} mn\overline{c^2}$, where m is the mass of a molecule, n is the number of molecules in the unit volume and $\overline{c^2}$ is the mean-square velocity (which is proportional to the absolute temperature).

(i) Thus, if n_1 be the number of molecules at a pressure p_1 and that n_2 at a pressure p_2, the temperature being the same, we have

$$\frac{n_1}{n_2} = \frac{p_1}{p_2}.$$

Here $n_1 = 2.7 \times 10^{19}$, $p_1 = 76$ cm,$p_2 = 10^{-6}$ mm $= 10^{-7}$ cm, $n_2 = ?$

$$\therefore \quad n_2 = \frac{n_1 \times p_2}{p_1}$$

$$= \frac{(2.7 \times 10^{19}) \times 10^{-7}}{76} = 3.55 \times 10^{10}.$$

(ii) Since $\overline{c^2} \propto = T$, we can write

$$p \propto nT$$

or $$n \propto p/T.$$

In this case, we have

$$\frac{n_1}{n_2} = \frac{p_1}{p_2} \times \frac{T_2}{T_1}$$

Here p_1 = 76 cm, T_1 = 0°C = 273 K,p_2 = 10^{-7} cm, T_2 = 39°C = 312 K.

$$\therefore \quad n_2 = \frac{n_1 \times p_2 \times T_1}{p_1 \times T_2}$$

$$= \frac{(2.7\times10^{19})\times10^{-7}\times273}{76\times312} = 3.11\times10^{10}.$$

2. ***Compute the number of molecules in 1 cu m of an ideal gas at 27°C and a pressure of 10 mm of mercury. The average translational kinetic energy of a molecule at 27°C is 6.2 × 10^{-21} joule; the density of mercury is 13.6 × 10^3 kg/cu m and g = 9.8 N/kg.***

Solution: The pressure exerted by an ideal gas is equal to two-third of the total translational kinetic energy of the molecules per unit volume of the gas. That is

$$p = \frac{2}{3} \text{ (K. E. per unit volume)}$$

or $$pV = \frac{2}{3} \text{K.E.}$$

or $$\text{K.E.} = \frac{3}{2} pV.$$

Here p = 10^{-2} metre of Hg = 10^{-2} × (13.6 × 10^3) × 9.8 = 1.33 × 10^3 newton /sq m and V = 1 cu m . Therefore, for the whole gas

$$\text{K. E.} = \frac{3}{2} \times (1.33 \times 10^3 \text{ newton/sq m}) \times 1 \text{ cu m}$$

$$= 2.0\times10^3 \text{ joule.}$$

The average kinetic energy per molecule is 6.2×10^{-21} joule.

$$\therefore \text{ no. of molecules } = \frac{2.0\times10^{3}}{6.2\times10^{-21}} = 3.2\times10^{23}$$

3. *If the density of nitrogen at N.T.P. is 1.25 gm/litre, calculate the root-mean-square speed of its molecules at 0°C and at 20°C.*

Solution: According to kinetic theory, the pressure p of a gas is given by,

$$P = \frac{1}{3}\, p\,\overline{c^2},$$

where p is the density of the gas and $\overline{c^2}$ the mean-square speed of its molecules. Therefore, the root-mean-square speed

$$c_{rms} = \sqrt{\overline{c^2}} = \sqrt{\frac{3p}{p}}$$

Here p = 1.013 × 10^5 newton/sq metre (normal) and p = 1.25 gm/litre = 1.25 kg/cu metre. Therefore, the rms speed at 0°C (normal) is

$$c_{rms} = \sqrt{\left(\frac{3\times(1.013\times10^{5})}{1.25}\right)} = 493 \text{ meter/sec.}$$

The rms speed is proportional to the square-root of the absolute temperature. Thus,

$$\frac{c_{1rms}}{c_{2rms}} \sqrt{\frac{T_1}{T_2}}$$

Here T_1 = 0°C = 273 K, T_2 = 20°C = 293 K, c_{1rms} = 493 metre/sec, c_{2rms} = ?

$$\therefore \quad c_{2rms} = c_{1rms} \times \sqrt{\frac{T_2}{T_1}}$$

$$= (493) \times \sqrt{\frac{293}{273}} = 511 \text{ metre/sec.}$$

4. ***Find the rms velocity of the molecules of nitrogen at 20°C. Molecular weight of nitrogen is 28 and R = 8.31 × 10^3 joule/(kilomole-degree).***

Solution: The rms velocity of the molecules of a gas of molecular weight M at temperature T is given by

$$c_{rms} = \sqrt{\frac{3RT}{M}}$$

Here M= 28 gm/mole = 28 kg/kilo-mole, R = 8.31 × 10^3 joule/(kilomole-K), T=20°C = 293K.

$$c_{rms} = \sqrt{\frac{3 \times 8.31 \times 10^3 \text{ joule/(kilo-mole K)} \times 293 \text{ K}}{28 \text{ kg/kilo-mole}}}$$

= 511 metre/sec.

5. ***Calculate the root-mean-square speea of the molecules of a gas which has a density of 1.24 × 10^{-2} kg/cu metre at a temperature of 273 K and a pressure of 1.00 × 10^{-2} atm. Identify the gas. (R = 8.31 joule/mole-K).***

Solution: $c_{rms} \sqrt{\frac{3p}{\rho}}$

Here p = 100 × 10^{-2} atm = 1.01 × 10^3 newton/sq metre. (∵ 1 atmos pressure = 1.01 × 10^5 newton/sq metre).

$$\therefore \quad C_{rms} = \sqrt{\frac{3 \times 1.01 \times 10^3}{1.24 \times 10^{-2}}} = 494 \text{ metre/sec.}$$

This is the root-mean-square speed of the gas molecules at 273 K. Now if M be the molecular weight of the gas, then

$$c_{rms} = \sqrt{\frac{3RT}{m}}$$

$$\therefore M = \frac{3RT}{(c_{rms})^2} = \frac{3\times 8.31\times 273}{(494)^2} = 0.0279\text{kg/mole}$$

$$= 27.9 \text{ g/mole.}$$

The molecular weight of nitrogen is 28. Hence, it is the diatomic gas nitrogen.

6. Calculate the root-mean-square speed of an argon atom at 27°C. At what temperature will the root-mean-square speed be (a) thrice, (b) half its value at 27°C ? The atomic weight of argon is 40.

Solution: $c_{rms} = \sqrt{\frac{3RT}{M}}$ R = 8.31 joule/mole-K, T= 27 + 273 = 300 K and M = 40 gm/mole= 40 × 10^{-3} kg/mole.

$$\therefore \quad c_{rms} = \sqrt{\frac{3\times 8.31\times 300}{40\times 10^{-3}}} = 432\text{metre/sec.}$$

(a)Let c_{rms} be the rms speed of argon at 300 K and $3c_{rms}$ at TK. The eq.

(i) shows that the mean-square-speed $(c_{rms})^2$ is proportional to the temperature. Thus,

$$\frac{(c_{rms})^2}{(3c_{rms})^2} = \frac{300}{T}$$

$$\therefore \quad T = 9\times 300 = 2700\text{K} = 2427°\text{C}.$$

(b) Again, let c_{rms} be the rms speed at 300 K and $\frac{1}{2}$ c_{rms} at T K. Then

$$\frac{(C_{rms})^2}{\left(\frac{1}{2}c_{rms}\right)^2} = \frac{300}{T}.$$

$$\therefore \quad T = \frac{1}{4} \times 300 = 75 \text{ K} = -198°\text{C}.$$

7. The root-mean-square speed of oxygen molecule is 460 metre per sec at 0°C. Calculate the rms speeds of helium and argon molecules at 40°C. The molecular weights of oxygen, argon and helium are 32, 40 and 4 gram/mole, respectively.

Solution: The root-mean-square speed c_{rms} for a gas is proportional to the square-root of the temperature (Kelvin) of the gas. Let us calculate its value for oxygen at 40°C.

$$\frac{(c_{rms})_{40}}{(3c_{rms})_0} = \sqrt{\frac{273+40}{273}} = \sqrt{\frac{313}{273}}.$$

$$\therefore \quad (C_{rms})_{40} = (C_{rms})_0 \times \sqrt{\frac{313}{273}}$$

$$= 460 \times \sqrt{\frac{313}{273}} = 492.5 \text{ meter/sec.}$$

Let us now compute the rms speeds of helium and argon at the same temperature. We know that

$$C_{rms} = \sqrt{\frac{3RT}{M}}.$$

$$\therefore \quad \frac{(C_{rms})_O}{(C_{rms})_{He}} = \sqrt{\frac{M_{He}}{M_O}}$$

$$\therefore \quad (C_{rms})_{He} = (C_{rms})_O \sqrt{\frac{M_O}{M_{He}}}$$

$$= 492.5 \times \sqrt{\frac{32}{4}} = 1393 \text{ metre/sec.}$$

Similarly, for argon

$$(C_{rms})_A = (C_{rms})_O \sqrt{\frac{M_O}{M_A}}$$

$$= 492.5 \times \sqrt{\frac{32}{40}}. = 440 \text{ metre/sec.}$$

8. ***At what temperature will the rms speed of hydrogen molecules be equal to the rms speed of nitrogen molecules at 35°C ? $M_N = 14\ M_H$.***

Solution. The rms velocity of the molecules of a gas of molecular weight M at Kelvin temperature T is given by

$$C_{rms} = \sqrt{\frac{3RT}{M}}.$$

Let M_N and M_H be the molecular weights of nitrogen and hydrogen; and T_N and T_H the corresponding Kelvin temperatures at which c_{rms} is same for both. Thus, we have

$$C_{rms} = \sqrt{\frac{3RT_N}{M_N}} = \sqrt{\frac{3RT_H}{M_H}}.$$

This gives

$$\frac{T_N}{M_N} = \frac{T_H}{M_H}.$$

Here $T_N = 35°C = 308\ K$ and $M_N/M_H = 14$.

$\therefore \quad T_H = T_N \times M_H/M_N = 308 \times 1/14$

$= 22K = -251°C.$

9. ***At what temperature will the rms speed of nitrogen molecules be half the rms speed of hydrogen molecules at 200 K ? Molecular weights of nitrogen and hydrogen are 28 and 2, respectively.***

Solution: $C_{rms} = \sqrt{\frac{3RT}{M}}.$

For hydrogen and nitrogen, we may write

$$C_{rms} = \sqrt{\frac{3RT_H}{M_H}}.$$

and $$\frac{1}{2} C_{rms} = \sqrt{\frac{3RT_N}{M_N}}.$$

These equations give

$$\tfrac{1}{4} = \frac{T_N}{M_N} \times \frac{M_H}{T_H}.$$

Here $T_H = 200$ K, $M_N/M_H = 28/2 = 14$, $T_N = ?$

$$\therefore \quad T_N = \tfrac{1}{4} T_H \times \frac{M_N}{M_H}$$

$$= \tfrac{1}{4} \times 200 \times 14 = 700 \text{ K}.$$

10. ***At what temperature will the average velocities of hydrogen and oxygen molecules be equal to the escape velocity from the surface of the earth ? Do you expect hydrogen and oxygen to be found in enough quantity in the earth's upper atmosphere where the temperature is about 10,000 K ? The escape velocity from earth's surface is 11.2 km/sec. Given: $k = 133 \times 10^{-23}$ joule/K, $N = 6.02 \times 10^{23}$ molecules/mole and molecular weights of hydrogen and oxygen are 2 and 32, respectively.***

Solution: According to kinetic theory, the average kinetic energy of a gas molecule at absolute temperature T is $\tfrac{3}{2}\, k\, T$, irrespective of its mass.

If v_e^2 be the escape velocity, then the kinetic energy required for the molecule to escape must be $\tfrac{1}{2}\, m\, v_c^2$, where m is the mass of the molecule.

Now, let m_H be the mass and T_H the temperature at which hydrogen molecule would escape, Then, we have

$$\tfrac{1}{2}\, m_H v_e^2 = \tfrac{3}{2}\, kT_H$$

or

$$T_H = \frac{m_H v_e^2}{3k}$$

Now, $m_H = \frac{2}{6.02 \times 10^{23}}\ g = 0.332 \times 10^{-26}$ kg and $v_e = 11.2 \times 10^3$ metre/sec.

$$\therefore \quad T_H = \frac{(0.332 \times 10^{-26})(11.2 \times 10^3)^2}{3 \times (1.38 \times 10^{-23})}$$

$$= 1.01 \times 10^4 \text{K} \simeq 10^4 \text{K}.$$

The molecular weight of oxygen is 32, that is, 16 times that of hydrogen, Hence, that corresponding temperature for oxygen is $T_O = 16 \times 10^4 \text{K} \simeq 10^5 \text{K}$.

The lighter gas hydrogen can attain enough speed by thermal agitation in the earth's upper atmosphere and may escape into outer space. Hence, hydrogen is not expected to be found in enough quantity in the earth's upper atmosphere. The heavier gas oxygen, however, would not attain enough speed so as to escape and is expected to be found in good quantity.

11. ***Given: Avogadro's number N ∓ 6.02 × 10^23 and Boltzmann's constant k = 1.38 × 10^-23 joule/molecule-K; calculate (a) the average kinetic energy of translation of the molecules of an ideal gas at 0°C and at 100°C. (b) Also calculate the corresponding energies per mole of the gas.***

Solution: (a) According to the kinetic theory, the average kinetic energy of translation per molecule of an ideal gas at Kelvin temperature T is $\frac{3}{2}\ kT$, where k is Boltzmann's constant.

At 0°C (T= 273 K), the kinetic energy of translation

$$= \frac{3}{2} kT$$

$$= \frac{3}{2} \times (1.38 \times 10^{-23}) \times 273 = 5.65 \times 10^{-21} \text{ joule/molecule.}$$

At 100°C (T= 373 K), the energy is

$$\frac{3}{2} \times (1.38 \times 10^{-23}) \times 373 = 7.72 \times 10^{-21} \text{ joule/molecule.}$$

(b) 1 mole of the gas contains $N (= 6.02 \times 10^{23})$ molecules. Therefore, at 0°C, the kinetic energy of translation of 1 mole of the gas is $(5.65 \times 10^{-21})(6.02 \times 10^{23}) \simeq$ 3400 joule/mole, and at 100°C, the kinetic energy of translation of 1 mole of gas is $(7.72 \times 10^{-21})(6.02 \times 10^{23}) \simeq$ 4647 joule/mole.

12. *The average kinetic energy of translation of a molecule of hydrogen at 0°C is 5.65 × 10⁻²¹ joule and the gas constant R is 8.31 joule/(mole-K). Calculate Avogadro's number.*

Solution: The average kinetic energy of translation per molecule of a gas is $\frac{3}{2} k\, T$, where k is Boltzmann's constant, or $\frac{3}{2} \frac{R}{N} T$, where N is Avogadro's number.

At 0°C (T=273 K), the translational energy per molecule is 5.65×10^{-21} joule and R = 8.31 joule/mole-K. Thus,

$$5.65 \times 10^{-21} = \frac{3}{2} \times \frac{8.31}{N} \times 273$$

$$\therefore \quad N = \frac{3}{2} \times \frac{8.31}{5.65 \times 10^{-21}} \times 273 = 6.02 \times 10^{23}.$$

13. *Calculate the molecular translational kinetic energy of a mole of a gas at 0°C. R = 8.31 joule/mole-K. Is it same for total kinetic energy of all gases ?*

Solution: The translational kinetic energy per molecule of an ideal gas at Kelvin temperature T is $\frac{3}{2} kT$. Since 1 mole of gas contains N (Avogadro's number) molecules, the translational kinetic energy per mole would be

$$\frac{3}{2} kT \times N = \frac{3}{2} RT. \qquad [\because k \times N = R]$$

Thus, for 1 mole of a gas at 0°C (= 273 K); the translational energy is $\frac{3}{2} \times 8.31 \times 273 = 3.4 \times 10^{3}$ joule

The translational kinetic energy per molecule (or per mole) will be same for all gases. The total kinetic energy will, however, be different for different gases, depending upon the atomicity (total degrees of freedom) of the gas.

14. ***Find the average translational kinetic energy of molecules of one mole of a gas contained in a volume 1.23 × 10⁻³ cu metre at a pressure 2 × 10⁵ newton/sq metre. Avogadro's number is 6.02 × 10²³ molecules/mole.***

Solution: The translational kinetic energy of one mole of a gas at kelvin temperature T is $\frac{3}{2}RT$, where R is gas constant. Since $pV = RT$ (for one mole), this can also be written as $\frac{3}{2}\ pV$. Here $p = 2 \times 10^5$ newton/sq metre and $V = 1.23 \times 10^{-3}$ cu metre. Therefore, the energy of one mole of the gas is

$$\frac{3}{2} \times (2 \times 10^5) \times (1.23 \times 10^{-3}) = 3.69 \times 10^2 \text{ joule.}$$

One mole of the gas has 6.02×10^{23} molecules. Therefore, the average translational kinetic energy per molecule is

$$\frac{3.69 \times 10^2}{6.02 \times 10^{23}} = 6.1 \times 10^{-22} \text{ joule.}$$

15. ***Given: Avogadro number $N = 6.02 \times 10^{23}$ and Boltzmann's constant $k = 1.38 \times 10^{23}$ joule/(molecule-K). Calculate (a) the average kinetic energy of translation of an oxygen molecule at 27°C, (b) the total kinetic energy of an oxygen molecule at 27°C, (c) the total kinetic energy of a gram-molecule of oxygen at 27°C.***

Solution: Kelvin temperature, $T = 27 + 273 = 300$ K.

We know from the law of equipartition of energy that the average kinetic energy of a gas molecule per degree of freedom is $\frac{1}{2}\ kT$.

(a) An oxygen molecule has 3 degrees of freedom with respect to translation. Hence, the average *kinetic energy of translation* of the molecule is

$$3 \times \frac{1}{2} kT = \frac{3}{2} kT$$

$$= \frac{3}{2} \times (1.38 \times 10^{-23}) \times 300$$

$$= 6.21 \times 10^{-21} \text{ joule/molecule.}$$

(b) An oxygen molecule, being diatomic, has 2 degrees of freedom with respect to rotation also, that is, a total of 5 degrees of freedom. Hence, the total kinetic energy of the molecule is

$$5 \times \frac{1}{2} kT = \frac{5}{2} kT$$

$$= \frac{5}{2} \times (1.38 \times 10^{-23}) \times 300$$

$$= 10.35 \times 10^{-21} \text{ joule/molecule.}$$

(c) One gm-molecule of oxygen contains N molecules. Hence, the total kinetic energy of 1 gm-molecule of the gas

$$= \text{N} \times \frac{5}{2} kT$$

$$= 6.02 \times 10^{23} \times 10.35 \times 10^{-21}$$

$$= 6231 \text{ Joule/mole.}$$

16. *Calculate the total kinetic energy of one kilogram-mole of oxygen at 27°C. The universal gas constant R is 8.31 × 10³ joule/kg-mole-K.*

Solution: According to the law of equipartition of energy, in a gas in equilibrium at kelvin temperature T, the average kinetic energy per molecule per degree of freedom is $\frac{1}{2} kT$, where k is Boltzmann's constant. An oxygen molecule, being diatomic, has 5 degrees of freedom (3 translational plus 2 rotational). Therefore, the total kinetic energy of the molecule is $\frac{5}{2} kT$.

Now, one kilogram-mole of oxygen has N molecules, where $N = R/k$, where R is universal gas constant in joule/kg-mole-K. Therefore, the total kinetic energy of one kg-mole of oxygen at temperature T (= 27°C = 300 K) is

$$= \frac{5}{2} kT \times \frac{R}{k} = \frac{5}{2} RT$$

$$= \frac{5}{2} (8.31 \times 10^3 \text{ joule/kg-mole-K}) \times 300 \text{ K}$$

$$= 6.23 \times 10^6 \text{ joule/kg-mole.}$$

17. *What is the energy of thermal motion of 20 gm of oxygen at 10°C ? What part of this energy falls to the share of translational motion and what of rotational motion ?*

Solution: An oxygen molecule has 5 degrees of freedom, 3 translational and 2 rotational. The total energy of thermal motion at 10°C per molecule, is therefore

$$= 5 \times \frac{1}{2} kT$$

$$= 5 \times \frac{1}{2} (1.38 \times 10^{-23} \text{ joule/K}) \times 283$$

K

$$= 9.76 \times 10^{-21} \text{ joule.}$$

The molecular weight of oxygen is 32, *i.e.*, the mass of 6.02×10^{23} molecules of oxygen is 32 gm. Therefore, the number of molecules in 20 gm of oxygen is $\frac{6.02 \times 10^{23}}{32} \times 20 = 3.76 \times 10^{23}$.

Hence, the total energy of 20 gm of gas is

$(9.76 \times 10^{-21}) \times (3.76 \times 10^{23}) = 3670$ joule.

The 3/5th part of this is translational and 2/3rd part is rotational.

18. *At what temperature will the kinetic energy of an atom be 1.0 eV ? (k = 1.38 × 10^{-23} joule/molecule-K).*

Solution: 10 eV (electron-volt) = 1.60×10^{-19} joule.

An atom consists of only 3 (translational) degrees of freedom. By the law of equipartition of energy, the average kinetic energy per degree of freedom at a Kelvin temperature T is $\frac{1}{2}\ kT$. Therefore, the average kinetic energy of the atom would be $\frac{3}{2}\ kT$. We want that

$$\frac{3}{2}\ kT = 1.60 \times 10^{-19} \text{ joule.}$$

$$\therefore \quad T = \frac{2}{3} \times \frac{1.60 \times 10^{-19}}{1.38 \times 10^{-23}} = 7729\text{K.}$$

19. ***The first excited state of hydrogen atom is 10.2 eV above its lowest state. At what temperature the hydrogen atom will be excited to this state ? k = 1.38 × 10⁻²³joule/K.***

Hence, explain why the gaseous atoms are excited to emit radiation by means of electric discharge rather than by thermal methods.

Solution: $1 \text{ eV} = 1.6 \times 10^{-19}$ joule.

$\therefore \quad 10.2 \text{ eV} = 16.32 \times 10^{-19}$joule.

Let T K be the temperature at which the hydrogen atom will have an average kinetic energy of 16.32×10^{-19} joule. Then, by the law of equipartition of energy, we have

$$\tfrac{3}{2}\ kT = 16.32 \times 10^{-19} \text{ joule.}$$

Now, the Boltzmann's constant $k = 1.38 \times 10^{-23}$ joule/K.

$$\therefore \quad \text{T} = \tfrac{2}{3}\ \frac{16.32 \times 10^{-19}}{k}$$

$$= \frac{2}{3} \times \frac{16.32 \times 10^{-19}}{1.38 \times 10^{-23}} \cong 8 \times 10^{4} \text{ K.}$$

Thus, we see that in order to excite the gaseous atoms by thermal method, a very high temperature is needed. At such temperature the atoms are excited by collision with other atoms, the necessary energy being derived from the kinetic energy of the colliding particles.

The excitation can, however, be comparatively easily achieved by means of electric discharge (arc, spark, Geissler tubes). In this method, electrons or ions are accelerated in an electric field and gain kinetic energy. When they collide with gas atoms, the atoms take the necessary energy from them and are excited. This method is much easier and hence preferred to thermal method to excite atoms.

20. ***The specific heat of a monatomic gas at constant volume, c_v is 0.75 cal/g-K. Calculate the atomic weight and mass of a gas atom. Given: R = 2 cal/mole-K and N = 6 × 10^{23}.***

Solution: Let M be the atomic weight of the gas (since the gas is monatomic, the atomic weight is same as the molecular weight.) Then the molar specific heat at constant volume is

$$C_V = Mc_V = 0.75\, M \text{ cal/mole-K.} \qquad [\because 1 \text{ mole} = M \text{ gm}]$$

By the law of equipartition of energy, the molar specific heat at constant volume, C_V, for a monatomic gas is $\frac{3}{2}$ R. Thus,

$$0.75 \frac{\text{cal}}{\text{mole-K}} \tfrac{3}{2} R = \tfrac{3}{2} \times 2 \frac{\text{cal}}{\text{mole-K}}$$

$$\therefore \qquad M = \frac{3}{0.75} = 4$$

One mole of a gas has 6 × 10^{23} molecules (or atoms in case of monatomic gas). Now, the mass of one mole of the gas is 4 gm. Therefore, the mass of one gas atom is

$$\frac{4}{6 \times 10^{23}} = 6.7 \times 10^{-24}\ g.$$

21. ***A cylinder-piston contains 32 kg of oxygen at 4 atmosphere pressure and 27°C temperature. It undergoes a cycle in which it is first heated at constant volume until the pressure becomes three times, then it is expanded isothermally until the pressure drops to the original value and finally it is cooled isobarically back to the original temperature.***

Calculate the change in internal energy and the work done in each process. Atomic wt of oxygen = 16 and R = 8.3 joule/mole-K.

Solution:

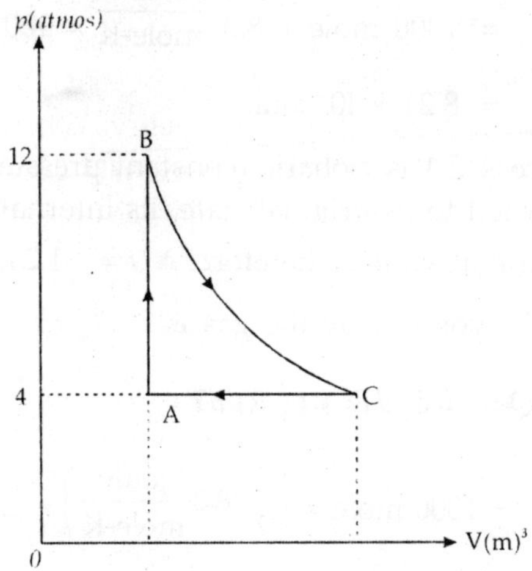

The cylinder contains 1000 moles of O_2 gas. The initial pressure is 4 atoms and temperature is 300 K.

The process *AB* occurs at constant volume, in which the pressure, and hence the temperature also, become three times, *i.e.*, 12 atoms and 900 K, respectively. At constant volume, the work done is zero, *i.e.*, *W = 0.*

Hence, by first law of thermodynamics, the change in the internal energy of the gas is

$$\Delta U = Q - W$$

$$= Q = \mu C_V \, dT = \mu \left(\tfrac{5}{2} R\right) dT$$

$$= 1000 \text{ mole} \left(\frac{5}{2} \times 8.3 \frac{\text{joule}}{\text{mole-K}}\right)(900 \text{ K} - 300 \text{ K})$$

$$= 1.245 \times 10^7 \text{ joule.}$$

The process *BC* is isothermal (constant temperature). Therefore $\Delta U = 0$.

The work done is given by

$$W = \mu R\,T \log (V_f V_i) = \mu RT \log_e 3$$

$$= 1000 \text{ mole} \times 8.3 \frac{\text{joule}}{\text{mole-K}} \times 900 \text{ K} \times 1.099$$

$$= 8.21 \times 10^6 \text{ joule.}$$

The process *CA* is isobaric (constant pressure). Since the gas has returned to its original state, its internal energy shall come to its initial value. Therefore $\Delta U = -1.245 \times 10^7$ joule.

The heat given out by the gas is

$$Q = \mu C_p\, dT = \mu\left(\tfrac{7}{2} R\right) dT$$

$$= 1000 \text{ mole} \times \left(\frac{7}{2} \times 8.3 \frac{\text{joule}}{\text{mole-K}}\right) \times (300 \text{ K} - 900 \text{ K})$$

$$= -1.743 \times 10^7 \text{ joule.}$$

Atomicity of Gas

In kinetic theory, the molecules of a gas are treated as hard, structureless particles. The actual picture of the molecules is, however, not so simple as a molecule is made up of one or more atoms. Larger the number of atoms in a molecule, the more complex it is. The number of atoms in a molecule of a gas determines the 'atomicity' of the gas. Those gases whose molecules consist of a single atom are called 'monatomic', such as mercury vapour, helium, argon, etc.; while those with two atoms in a molecule are called' diatomic', such as oxygen, hydrogen, etc. The gases having more than two atoms in a molecule are called 'polyatomic', such as carbon dioxide.

It is now to be seen whether the atomicity of the gas affects the derivations of the kinetic theory. In this theory, it is assumed that the collisions between gas molecules are perfectly elastic, that is, the sum of the kinetic energies of translation of two molecules is the same before and after the collision. In other words, the kinetic energy of translation is conserved. The assumption clearly holds for a monatomic gas whose molecules have only the kinetic energy of translation. The molecules of a diatomic (or polyatomic) gas, however, not only move bodily but also rotate about an axis. Thus, they possess kinetic energy of translation as well as of rotation. Also, the atoms within a diatomic or polyatomic molecule may be vibrating with respect to one another. If this is the case, the molecules would possess energy of vibration also.

Thus, in a collision between two polyatomic molecules there can be an exchange between the kinetic energy of translation and the energy of rotation and vibration. As such the kinetic energy of translation may not be conserved in an individual collision, but it must be conserved on the whole because the results obtained from this assumption hold for monatomic as well as for polyatomic gases.

Since a molecule undergoes a large number of collisions, we can suppose that the change of translational energy into rotational and vibrational energies is balanced by the change of rotational and vibrational energies into translational energy.

Thus, on the whole, the translational energy is conserved. Hence, the results obtained from kinetic theory must be true for polyatomic gases also.

Relation between γ ***of a Gas and its Atomicity:*** Let us consider one mole of a perfect gas whose temperature increases by dT on being heated. The increase in the *total* energy (translational + rotational + vibrational) of the gas is $C_V dT$ where C_v is the molecular specific heat of the gas at constant volume. Now, we know that the average kinetic energy of translation of a single molecule is $\frac{3}{2}\,kT$, where k is Boltzmann's constant.

Therefore, the increase in the transnational energy of a mole of the gas will be

$$= \tfrac{3}{2} k\, dT \times N, \text{ where } N \text{ is Avogadro's number}$$

$$= \tfrac{3}{2} RdT. (\because\ k = R/N)$$

where R is the universal gas constant.

Hence,

$$\frac{\text{increase in total energy}}{\text{increase in translational energy}} = \beta = \frac{C_V dT}{\tfrac{3}{2} RdT} = \frac{2}{3}\frac{C_V}{R}.$$

But $R = C_p - C_v$ (Mayer's relation).

$$\beta = \frac{2}{3}\frac{C_V}{Cp - C_V} = \frac{2}{3}\frac{1}{(C_p/C_V) - 1} = \frac{2}{3}\frac{1}{\gamma - 1}$$

where $\gamma = C_p/C_V$. This equation determines the relation between y of a gas and its atomicity. For gases like argon, helium or mercury vapour, the experimental value of y comes out to be $\frac{5}{3}$. Therefore, for these gases

$$\beta = \frac{2}{3}\left(\frac{1}{\frac{5}{3} - 1}\right) = 1$$

that is, the total energy is the same as the translational energy. Since this is true for monatomic gases, so all those gases for which $\gamma = \frac{5}{3}$ are monatomic. That the mercury vapour is monatomic has been confirmed chemically. For gases like oxygen, nitrogen, the value of γ comes out to be $\frac{7}{3}$. Therefore

$$\beta = \frac{2}{3}\left(\frac{1}{\frac{7}{5} - 1}\right) = \frac{5}{3}$$

This means that the (rotational + vibrational) energy is two-third of the translational energy. Thus, we find that smaller the value of γ for a gas, greater is the value of β, that is, greater is the (rotational + vibrational) energy than the translational energy. Hence, we conclude that smaller the value of γ, greater is the atomicity of the gas.

Significance of Kinetic Theory

The kinetic theory of matter has been built on two basic facts about matter and heat: (i) "matter is discontinuous in structure" and (ii) "heat is a form of energy." It may be summarised as below:

(i) *Matter is Discontinuous, being Composed of a Very Large Number of Tiny Particles, Called Molecules:* These molecules retain the characteristic properties of the substance which they compose.

(ii) *The Molecules of Matter are Constantly in a State of Rapid Rectilinear Motion:* The diffusion of gases into one another, the free evaporation from liquid surfaces, and the diffusion that takes place at the common surface of two solids pressed together for considerable period indicate that the molecular motion is associated with all the three forms of matter. In liquids, however, the motion of the molecules is restricted than in gases, and even more restricted in solids.

(iii) *The Energy of the Molecules of Matter Constitutes the Heat Content of the Matter:* In general, the heat content is partly in the form of kinetic energy and partly in the form of potential energy of the molecules of matter.

Kinetic theory explains the three states of matter in the following way:

The Solid State: The molecules of a solid are so densely packed together that they exert considerable force of attraction on one another. Therefore, they cannot move freely throughout the substance, but can simply oscillate about their fixed positions of equilibrium. This is why the solids have a definite size and a definite shape. When a solid is heated, its molecules gain energy and the distance between their equilibrium positions increases, causing the body to expand.

This is thermal expansion. On further heating, a stage is reached when the molecules become so much apart that the forces of attraction become very feeble. The molecules then move freely in the body of the solid which therefore takes the form of liquid.

This is melting of solid. Now, the molecules in a liquid are separated from one another much more than in a solid at the same temperature. Therefore, the heat taken in by the solid, when melting, is used up in pulling the molecules apart against their mutual attraction. Hence, the temperature does not rise. This is the patent heat of fusion of the solid.

The Liquid State: The molecules of a liquid are less closer to one another that those of a solid. Therefore, they exert a smaller attraction on one another. This attraction, however, does not allow the molecules to leave the liquid but the molecules can move freely in the interior of the liquid. The liquid, therefore, adjusts itself to the shape of the containing vessel. This is why a liquid has a definite size, but not a definite shape.

A molecule in the interior of the liquid is attracted equally in all directions by the surrounding molecules and hence it experiences no resultant force of attraction. The molecules near the surface of the liquid, however, are acted upon by a mean force of attraction due to the molecules below it. This force, which is directed towards the interior of the liquid, causes the surface to behave like a stretched membrane. This is the explanation of surface tension.

Cooling vs Evaporation

All the molecules of a liquid do not move with the same speed, but the speeds are distributed about an average value which depends on the temperature of the liquid and increases as the temperature rises. Those molecules of the surface whose speeds are above the average value overcome the attraction of the other molecules in the surface and escape.

This is evaporation. The average speed of the remaining molecules drops correspondingly and hence the temperature of the liquid falls. This is why evaporation causes cooling of the liquid.

Practical Fields

Summing Up of Infinite Series

(a) *Prove* $\frac{\pi^2}{6} = 1 + \frac{1}{2^2} + \frac{1}{3^2} + \frac{1}{4^2} + \ldots$ ...(i)

Let $f(x) = x^2, -\pi < x < \pi$

This is an even function and therefore all the sine terms in its Fourier expansion are zero

So $f(x) = a_0 + \sum_{n=1}^{\infty} a_n \cos nx,$

where $a_0 = \frac{1}{2\pi} \int_{-\pi}^{\pi} x^2 dx = \frac{\pi^2}{3}$ ...(ii)

$$a_n = \frac{2}{\pi} \int_0^{\pi} x^2 \cos n\, x\, dx = (-1)^n \frac{4}{n^2} \quad \ldots\text{(iii)}$$

Thus, $x^2 = \frac{\pi^2}{3} + 4\left[-\cos x + \frac{\cos 2x}{4} - \frac{\cos 3x}{9} + \ldots\right]$ ...(iv)

For $x = \pi$, this reduces to

$$x^2 = \frac{\pi^2}{3} + 4\left[1 + \frac{1}{4} + \frac{1}{9} + \ldots\right]$$

or $$\frac{\pi^2}{6} = \frac{1}{1} + \frac{1}{2^2} + \frac{1}{3^2} + \ldots \qquad \ldots(v)$$

Hence, the result.

(b) *Determine the Fourier series expansion of the Function*

$$f(x) = \pi^2 - x^2 \qquad \text{for} \quad -\pi \le x \le \pi$$

Here the time period is 2π.

The Fourier series will be given as

$$\frac{A_0}{0} + \sum_{n=1}^{\infty}\left[A_n \cos(nx) + B_n \sin(nx)\right]$$

where the Fourier coefficients A_0 is given by

$$A_0 = \frac{1}{\pi}\int_{-n}^{n}\left(\pi^2 - x^2\right)dx = \frac{4\pi^2}{3}$$

and $$A_n = \frac{1}{\pi}\int_{-n}^{n}\left(\pi^2 - x^2\right)\cos(nx)\,dx \qquad \text{for } n = 1, 2, \ldots$$

$$= \frac{1}{\pi}\left\{\int_{-n}^{0}\left(\pi^2 - x^2\right)\cos(nx)\,dx + \int_{0}^{\pi}\left(\pi^2 - x^2\right)\cos(nx)\,dx\right\}$$

(Change $x \to -x$)

$$= \frac{1}{\pi}\left\{-\int_{-n}^{0}\left(\pi^2 - x^2\right)\cos(-nx)\,dx + \int_{0}^{\pi}\left(\pi^2 - x^2\right)\cos(nx)\,dx\right\}$$

$$= \frac{1}{\pi}\left\{\int_{0}^{\pi}\left(\pi^2 - x^2\right)\cos(nx)\,dx + \int_{0}^{\pi}\left(\pi^2 - x^2\right)\cos(nx)\,dx\right\}$$

$$= \frac{2}{\pi}\int_{0}^{\pi}\left(\pi^2 - x^2\right)\cos(nx)\,dx$$

Integrating by parts twice we get

$$A_n = \frac{2}{\pi}\left(\pi^2 - x^2\right)\frac{\sin(nx)}{n}\Bigg|_0^{\pi} - \frac{2\pi}{n}\int_{0}^{\pi}(-2x)\sin(nx)\,dx$$

$$= 0 + \frac{4}{\pi n}\frac{x(-\cos nx)}{n}\Bigg|_0^{\pi} + \frac{4}{\pi n^2}\int_0^{\pi} 1.\cos(nx)\,dx$$

$$= -\frac{4\pi}{\pi n^2}\cos(n\pi) + \frac{4}{\pi n^3}\sin(nx)\Big|_0^{\pi} + \frac{4}{n^2}(-1)^{n+1} + 0$$

$$\therefore \quad A_n = \frac{4(-1)^{n+1}}{n^2} \quad \text{for } n = 0, 1, 2, ...$$

Now,

$$B_n = \frac{1}{\pi}\int_{-\pi}^{\pi}(\pi^2 - x^2)\sin(nx)\,dx \quad \text{for } n = 1, 2, ...$$

Changing $x \to -x$, we have

$$B_n = \frac{1}{\pi}\int^{-\pi}\left(\pi^2 - (-x)^2\right)\sin(-nx)(-dx)$$

$$= +\frac{1}{\pi}\int_{\pi}^{-\pi}(\pi^2 - x^2)\sin(nx)(dx)$$

$$= -\frac{1}{\pi}\int_{-\pi}^{-\pi}(\pi^2 - x^2)\sin(nx)(dx)$$

$$= -B_n$$

so $\quad 2B_n = 0$

or $\quad B_n = 0$

Thus, the series is

$$f(x) = \frac{2\pi^2}{3} + 4\left(\cos x - \frac{1}{2^2}\cos 2x + \frac{1}{3^2}\cos 3x + ...\right)$$

putting $x = 0$, we have

$$\pi^2 = \frac{2\pi^2}{3} + 4\left(1 - \frac{1}{2^2} + \frac{1}{3^2} - \frac{1}{4^2} + ...\right)$$

or $$\frac{\pi^2}{12} = 1 - \frac{1}{2^2} + \frac{1}{3^2} - \frac{1}{4^2} + \frac{1}{5^2} - ... \qquad ...(1)$$

Again putting $x = \pm\pi$

$$0=\frac{2\pi^2}{3}+4\left(\cos\pi-\frac{1}{2^2}\cos 2\pi+\frac{1}{3^2}\cos 3\pi...\right)$$

or
$$0=\frac{2\pi^2}{3}-4\left(1+\frac{1}{2^2}+\frac{1}{3^2}-\frac{1}{4^2}+...\right)$$

$\therefore$
$$\frac{\pi^2}{6}=1+\frac{1}{2^2}+\frac{1}{3^2}+\frac{1}{4^2}+... \qquad ...(2)$$

Adding (1) and (2), we have

$$\frac{\pi^2}{4}=2+\frac{2}{3^2}+\frac{2}{5^2}+...$$

or
$$\frac{\pi^2}{8}=1+\frac{1}{3^2}+\frac{1}{5^2}+\frac{1}{7^2}+...$$

(c) *Determine the Fourier series expansion of the function* $f(x)=|x|$ for $-\pi\le x\le\pi$.

Hence, show that $\dfrac{\pi^2}{8}=1+\dfrac{1}{3^2}+\dfrac{1}{5^2}+\dfrac{1}{7^2}+...$

Now $f(-x)=|-x|=|x|=f(x)$ and so is even.

Therefore it can be expanded only in cosine series and the coefficients $A_0, A_1, A_2,....$ can be evaluated from half range only, i.e. the Fourier series is

$$f(x)=\frac{A_0}{2}+\sum_{n=1}^{\infty}A_n\cos(nx)$$

where,

$$A_0=\frac{2}{\pi}\int_0^{\pi}|x|\,dx=\frac{2}{\pi}\int_0^{\pi}x\,dx$$

$$=\frac{2}{\pi}\left.\frac{x^2}{2}\right|_0^{\pi}=\frac{\pi^2}{\pi}=\pi$$

$$A_n = \frac{2}{\pi}\int_0^{\pi} |x| \cos(nx)\,dx$$

$$= \frac{2}{\pi}\int_0^{\pi} x\cos(nx)\,dx$$

$$= \frac{2}{\pi}\frac{x\sin(nx)}{(-n)}\bigg|_0^{\pi} - \frac{2}{\pi n}\int_0^{\pi}(1)\sin(nx)\,dx$$

$$= -\frac{2}{\pi n}\frac{\cos(nx)}{(-n)}\bigg|_0^{\pi} = \frac{2}{n^2}\left[(-1)^n - 1\right]$$

$$= \frac{2}{\pi n^2}\left[(-1)^n - 1\right] \text{ for } n = 1, 2, 3, \ldots$$

$$\therefore \quad A_1 = -\frac{4}{1^2\pi},\ A_2 = \frac{1}{2^2}(1-1) = 0$$

$$A_3 = -\frac{4}{3^2\pi},\ A_4 = 0,\ A_5 = -\frac{4}{5^2\pi},\ \ldots$$

$$|x| = \frac{\pi}{2} - \frac{4}{1^2\pi}\cos x - \frac{4}{\pi 3^2}\cos 3x - \frac{4}{5^2\pi}\cos 5x - \ldots$$

for $x = \pm\pi$, we have

$$\pi = \frac{\pi}{2} + \frac{4}{\pi}\left(1 + \frac{1}{3^2} + \frac{1}{5^2} + \frac{1}{7^2} + \ldots\right)$$

$$\therefore \quad \frac{\pi^2}{8} = 1 + \frac{1}{3^2} + \frac{1}{5^2} + \frac{1}{7^2} + \ldots$$

Applications, Which are Physical

Analysis of Fourier Series

(a) *Analysis of a Square Wave by Foureirs Theorem*

Let us consider any physical quantity y (such as displacement, pressure or electric current) varying periodically as shown in Fig. Here y has a constant value *a* from time *t=0* to $t=T/2$ and then remains zero for the remainder of its period T. We may express this function as follows:

$$\left.\begin{aligned} y &= a \text{ from } t = 0t = T/2 \\ \text{and } y &= 0 \text{ from } t = 0t = T/2 \end{aligned}\right\} \qquad \text{...(i)}$$

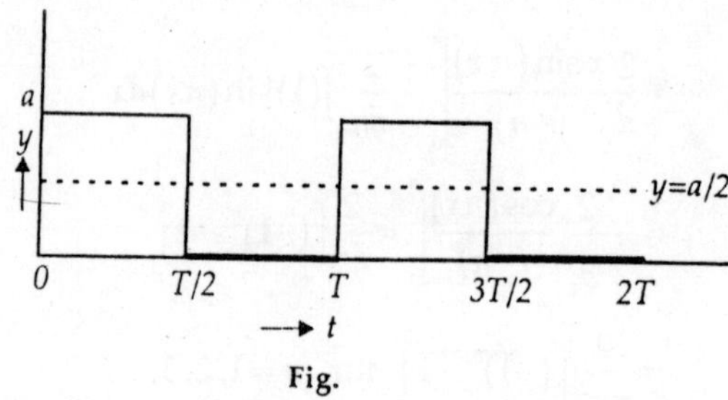

Fig.

when the time-axis of coordinates is taken through the lowest point of the displacement curve.

The Fourier's theorem can be expressed as

$$y = f\ (t) = A_0 + A_1 \cos \omega t + \ldots + A_r \cos r\omega t + \ldots$$

$$+B_1 \sin \omega t + \ldots + B_r \sin r\omega_t + \ldots \qquad \text{...(ii)}$$

where $A_0 = \dfrac{1}{T}\int\limits_0^T y dt, A_r = \dfrac{2}{T}\int\limits_0^T y \cos r\omega t\ dt$

and $B_r = \dfrac{2}{T}\int\limits_0^T y \sin r\omega t\ dt$

Let us obtain the values of A_0, A_r and B_r for the given function:

$$A_0 = \frac{1}{T}\int\limits_0^T y dt.$$

Substituting the value of y from eq. (i) we get

$$A_0 = \frac{1}{T}\int\limits_0^{T/2} a\ dt.$$

$$= \frac{a}{T}[t]_0^{T/2} = \frac{a}{2},$$

i.e., the axis of the displacement curve is the line $y = \frac{a}{2}$ (shown dotted in fig.).

For A_r, we have

$$A_r = \frac{2}{T}\int_0^T y \cos r\omega t \, dt$$

$$= \frac{2}{T}\int_0^T y \cos \frac{2r\pi t}{T} \, dt. \quad \left[\because \omega = \frac{2\pi}{T}\right]$$

Substituting the value of y from eq. (i), we have

$$A_r = \frac{2}{T}\int_0^{T/2} a \cos \frac{2r\pi t}{T} \, dt$$

$$= \frac{2a}{T}\frac{T}{2r\pi}\left[\sin \frac{2r\pi t}{T}\right]_0^{T/2}$$

$$= \frac{a}{r\pi}[\sin r\pi - \sin 0]$$

$$= 0.$$

Hence, all the cosine terms in eq. (ii) are zero.

Similarly, for b_r, we have

$$B_r = \frac{2}{T}\int_0^{T/2} y \sin r\omega t \, dt$$

$$= \frac{2}{T}\int_0^{T/2} y \sin \frac{2r\pi t}{T} \, dt.$$

Substituting the value of y from eq. (i) we have

$$B_r = \frac{2}{T}\int_0^{T/2} a \sin \frac{2r\pi t}{T} \, dt$$

$$= \frac{2a}{T}\frac{T}{2r\pi}\left[-\cos\frac{2r\pi t}{T}dt\right]_0^{T/2}$$

$$= \frac{a}{r\pi}[-\cos r\pi - \cos 0]$$

$$= \frac{a}{r\pi}[-\cos r\pi + 1]$$

when r is even, $\cos r\pi = +1$ and when r is odd, $\cos r\pi = -1$. Thus,

$$B_r = \frac{a}{r\pi}\left[-\begin{Bmatrix} 1 \text{ for even } r \\ -1 \text{ for odd } r \end{Bmatrix} + 1\right]$$

$$= \frac{a}{r\pi}\left[(-1)^r + 1\right] = \frac{a}{r\pi}\left[1 + (-1)^{r+1}\right].$$

Putting $r = 1, 2, 3, \ldots$, we get

$$B1 = \frac{2a}{\pi} B_2 = 0, B_3 = \frac{2a}{3\pi}, B_4 = 0, B_5 = \frac{2a}{5\pi}, \ldots$$

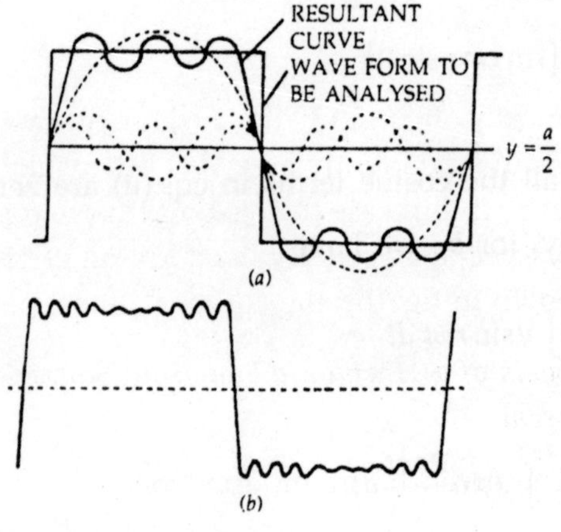

Fig.

Therefore, putting the values of A_0, A_r and B_r, eq. (ii) becomes

$$y = \frac{a}{2} + \frac{2a}{\pi}\left[\sin \omega t + \frac{1}{3}\sin 3\omega t + \frac{1}{5}\sin 5\omega t + \ldots\right]$$

$$\text{or } y = \frac{a}{2} + \frac{2a}{\pi}\left[\sin\frac{2\pi t}{T} + \frac{1}{3}\sin\frac{6\pi t}{T} + \frac{1}{5}\sin\frac{10\pi t}{T} + \ldots\right]$$

Thus, the given complex motion has the axis $y = a/2$ and the various components as under:

$$\frac{2a}{\pi}\sin\frac{2\pi t}{T}$$

$$\frac{2a}{3\pi}\sin\frac{6\pi t}{T}$$

$$\frac{2a}{5\pi}\sin\frac{10\pi t}{T}, \text{and so on,}$$

having frequencies in the ratio 1: 3: 5, etc. and amplitudes in the ratio

$$1:\frac{1}{3}:\frac{1}{5} \text{ etc.}$$

In Fig., the addition of the successive terms is indicated graphically. In fig. (a) the first three terms of the series are shown independently and also the resultant curve. It is seen that even in this case the resultant curve roughly resembles the curve to be analysed. By the addition of 15 terms of the series as shown in fig. (b); the curve is almost perfectly reproduced.

Analysis of a Harmonic Function (Square Wave) by Fourier's Theorem

The given harmonic function

$$\left.\begin{aligned} &y = a \text{ from } t = 0 \textit{ to } t = T/2 \\ \text{and } &y = -a \text{ from } t = T/2 \textit{ to } t = T \end{aligned}\right\} \quad \ldots\text{(i)}$$

represents a square wave as shown in fig.

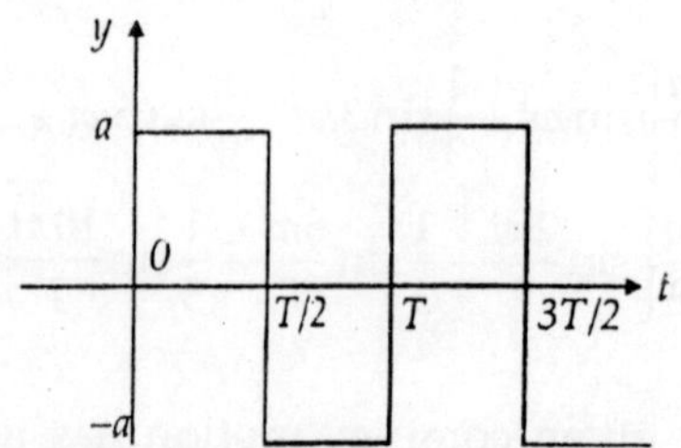

The Fourier series of a harmonic function is

$$y = f(t) = A_0 + A_1 \cos \omega t + \ldots + A_r \cos r\omega\, t + \ldots$$
$$= B_1 \sin \omega t + \ldots + B_r \sin r\omega t + \ldots,$$

where $A_0 = \frac{1}{T}\int_0^T y\,dt, A_r = \frac{2}{T}\int_0^T y \cos r\omega\, t\, dt$

and $B_r = \frac{2}{T}\int_0^T y \sin r\omega\, t\, dt.$

Let us obtain the values of A_0, A_r and B_r for the given function:

$$A_0 = \frac{1}{T}\int_0^T y\, dt$$

Substituting the values of y from eq. (i), we get

$$A_0 = \frac{1}{T}\int_0^T y\, dt - \frac{1}{T}\int_{T/2}^T a\, dt$$
$$= \frac{a}{T}[t]_{T/2}^T$$

Thus, the axis of the displacement curve is the x-axis (y=0). For A_r, we have

$$A_r = \frac{2}{T}\int_0^T y \cos(r\omega t)t$$
$$= \frac{2}{T}\int_0^T y \cos\frac{2r\pi t}{T}dt. \qquad \left[\because w = \frac{2\pi}{T}\right]$$

Substituting the value of y from eq. (i), we get

$$Ar = \frac{2}{T}\int_0^{T/2} a \cos\frac{2r\pi t}{T}dt - \frac{2}{T}\int_{T/2}^T a \cos\frac{2r\pi t}{T}dt$$

$$= \frac{2a}{T}\frac{T}{2r\pi}\left[\sin\frac{2r\pi t}{T}\right]_0^{T/2} - \frac{2a}{T}\frac{T}{2r\pi}\left[\sin\frac{2r\pi t}{T}\right]_{T/2}^{T}$$

$$= \frac{a}{r\pi}[\sin r\pi - \sin 0] - \frac{a}{r\pi}[\sin 2r\pi - \sin r\pi]$$

$$= \frac{a}{r\pi}[2\sin r\pi - \sin 2r\pi] = 0.$$

Hence, all the cosine term in eq. (ii) are zero. For B_r, we have

$$B_r = \frac{2}{T}\int_0^t y\sin(r\omega t)\,dt$$

$$= \frac{2}{T}\int_0^T y\sin\frac{2r\pi t}{T}dt$$

$$= \frac{2}{T}\int_0^{T/2} a\sin\frac{2r\pi t}{T}dt - \frac{2}{T}\int_{T/2}^{T} a\sin\frac{2r\pi t}{T}dt$$

$$= \frac{2a}{T}\frac{T}{2r\pi}\left[-\cos\frac{2r\pi t}{T}\right]_0^{T/2} - \frac{2a}{T}\frac{T}{2r\pi}\left[-\cos\frac{2r\pi t}{T}\right]_{T/2}^{T}$$

$$= \frac{a}{r\pi}[-\cos r\pi + \cos 0] - \frac{a}{r\pi}[-\cos 2r\pi + \cos r\pi]$$

$$= \frac{a}{r\pi}[-\cos r\pi + 1 + \cos 2r\pi - \cos r\pi]$$

$$= \frac{a}{r\pi}[2 - 2\cos r\pi] \qquad [\because \cos 2r\pi = 1]$$

$$= \frac{2a}{r\pi}[1 - \cos r\pi].$$

But cos $r\pi = (-1)^r$, where r = 1,2,3,...

$$\therefore B_r = \frac{2a}{r\pi}\left[1 - (-1)^r\right].$$

Putting r = 1, 2, 3,...we get

$$B_1 = \frac{4a}{\pi}, B_2 = 0, B_3 = \frac{4a}{3\pi}, B_4 = 0 B_5 = \frac{4a}{5a}, \ldots$$

Putting the values of A_0, A_r, and B_r, in eq. (ii), we get

$$y = \frac{4a}{\pi}\left[\sin\omega t + \frac{1}{3}\sin 3\omega t + \frac{1}{5}\sin 5\omega t + \ldots\right]$$

$$\text{or } y=\frac{4a}{\pi}\left[\sin\frac{2\pi t}{T}+\frac{1}{3}\sin\frac{2\pi t}{T}+\frac{1}{5}\sin\frac{10\pi t}{T}+\ldots\right]$$

Thus, the given function can be written as a sum of simple harmonic components having frequencies in the ratio 1 : 3 : 5 :.....and amplitudes in the ratio $1:\frac{1}{3}:\frac{1}{5}:\ldots$

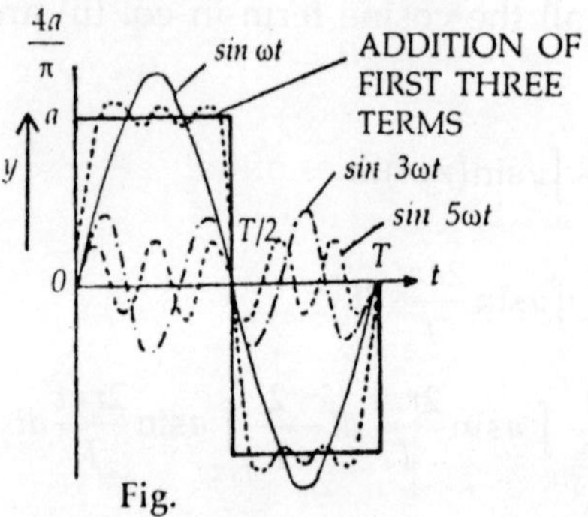

Fig.

If we take the first three terms of the above series representing the square wave and add them together, the result is Fig. The First harmonic has the frequency of the square wave and the higher frequencies build up the squareness of the wave. The highest frequencies are responsible for the sharpness of the vertical sides of the wave.

(b) *Analysis of a Saw-Tooth Wave by Fourier's Theorem*

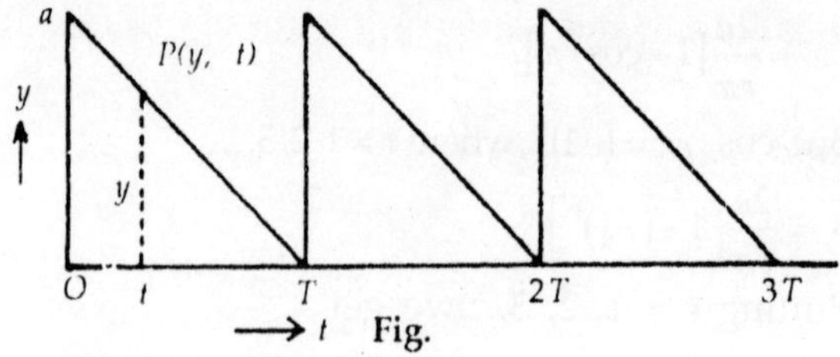

Fig.

The curve shown in fig. represents a saw-tooth wave. It can be expressed mathematically as

$$y=f(t)=a\left(1-\frac{t}{T}\right),\text{ from } t=0 \text{ to } t=T. \qquad \ldots\text{(i)}$$

The Fourier's series is

$$y = f(t) = A_0 + A_1 \cos \omega t + ... + A_r \cos r\omega t + ... \qquad ...(ii)$$
$$+ B_1 \sin \omega t + ... + B_r \sin r\omega t + ...$$

where $A_0 = \frac{1}{T}\int_0^T y\, dt, A_r = \frac{2}{T}\int_0^T y \cos r\,\omega t\, dt$

and $B_r = \frac{2}{T}\int_0^T y \sin r\,\omega t\, dt.$

Let us obtain the values of the coefficients A_0, A_r and B_r of the Fourier series:

$$A_0 = \frac{1}{T}\int_0^T y\,dt = \frac{a}{T}\int_0^T \left(1 - \frac{t}{T}\right) dt$$

$$= \frac{a}{T}\left[t - \frac{t^2}{2T}\right]_0^T = \frac{a}{T}\left[T - \frac{T^2}{2T}\right] = \frac{a}{2}.$$

which gives the ordinate of the axis of the curve.

For A_r, we have

$$A_r = \frac{2}{T}\int_0^T y \cos r\,\omega\, t\, dt$$

$$= \frac{2}{T}\int_0^T y \cos \frac{2r\pi t}{T} dt. \qquad \left[\because \omega = \frac{2\pi}{T}\right]$$

Substituting the value of y from eq. (i), we have

$$A_r = \frac{2}{T}\int_0^T a\left(1 - \frac{t}{T}\right) \cos \frac{2r\pi t}{T} dr$$

$$= \frac{2a}{T}\int_0^T \cos \frac{2r\pi t}{T} dt - \frac{2a}{T^2}\int_0^T t \cos \frac{2r\pi t}{T} dt$$

$$= -\frac{2a}{T^2}\int_0^T t \cos \frac{2r\pi t}{T} dt \qquad \left[\because \int_0^T \cos \frac{2r\pi t}{T} dt = 0\right]$$

$$= -\frac{2a}{T^2}\left[t \frac{\sin(2r\pi t/T)}{2r\pi/T} - \int \frac{\sin(2r\pi t/T)}{2r\pi/T}\right]_0^T \qquad \text{(by parts)}$$

$$= -\frac{2a}{T^2}\left[t\frac{\sin(2r\pi t/T)}{2r\pi/T} + \frac{\cos(2r\pi t/T)}{(2r\pi/T)^2}\right]_0^T$$

$$= -\frac{2a}{T^2}\left[T\frac{\sin 2r\pi}{2r\pi/T} + \frac{\cos 2r\pi}{(2r\pi/T)^2} - 0 - \frac{\cos 0}{(2r\pi/T)^2}\right]$$

$$= -\frac{2a}{T^2}\left[\frac{1}{(2r\pi/T)^2} - \frac{1}{(2r\pi/T)^2}\right]$$

$$= 0. \qquad [\because \sin 2r\pi = 0 \text{ and } \cos 2r\pi = 1]$$

For B_r, we have

$$B_r = \frac{2}{T}\int_0^T y \sin r\omega t \, dt$$

$$= \frac{2}{T}\int_0^T y \sin\frac{2r\pi t}{T} dt. \qquad \left[\because \omega = \frac{2\pi}{T}\right]$$

Substituting the value of y from eq. (i), we get

$$B_r = \frac{2}{T}\int_0^T a\left(1 - \frac{t}{T}\right)\sin\frac{2r\pi t}{T} dt$$

$$= \frac{2a}{T}\int_0^T \sin\frac{2r\pi t}{T} dt - \frac{2a}{T^2}\int_0^T t\sin\frac{2r\pi t}{T} dt$$

$$= \frac{2a}{T^2}\int_0^T t\sin\frac{2r\pi t}{T} dt \qquad \left[\because \int_0^T \sin\frac{2r\pi t}{T} dt = 0\right]$$

$$= \frac{2a}{T^2}\left[t - \frac{\cos(2r\pi t/T)}{2r\pi/T} - \int\frac{-\cos(2r\pi t/T)}{2r\pi/T}\right]_0^T \quad \text{(by parts)}$$

$$= \frac{2a}{T^2}\left[t\frac{\cos(2r\pi t/T)}{2r\pi/T} - \frac{\sin(2r\pi t/T)}{(2r\pi/T)^2}\right]_0^T$$

$$= \frac{2a}{T^2}\left[t\frac{\cos 2r\pi}{2r\pi/T} - \frac{\sin 2r\pi}{(2r\pi/T)^2} - 0 + \frac{\sin 0}{(2r\pi/T)^2}\right]$$

$$= \frac{2a}{T^2}\left[T\frac{1}{2r\pi/T}\right] \quad [\because \cos 2r\pi = 1 \text{ and } \sin 2r\pi = 0]$$

$$= \frac{a}{r\pi}.$$

Thus $B_1 = \frac{a}{\pi}, B_2 = \frac{a}{2\pi}, B_3 = \frac{a}{3\pi}, ...$

Substituting the values of A_0, A_r and B_r in eq. (ii) we have

$$y = \frac{a}{2} + \frac{a}{\pi}\left(\sin \omega t + \frac{1}{2}\sin 2\omega t + \frac{1}{3}\sin 3\omega t + ...\right)$$

or $$y = \frac{a}{2} + \frac{a}{\pi}\left(\sin\frac{2\pi t}{T} + \frac{1}{2}\sin\frac{4\pi t}{T} + \frac{1}{3}\sin\frac{6\pi t}{T} + ...\right).$$

Thus, the given periodic motion has the axis $y = \frac{a}{2}$ above the time-axis, and the various components as under

$$\frac{a}{\pi}\sin\frac{2\pi t}{T},$$

$$\frac{a}{2\pi}\sin\frac{4\pi t}{T},$$

$$\frac{a}{3\pi}\sin\frac{2\pi t}{T},$$

and so on, having frequencies in the ratio 1 : 2 : 3... and amplitudes in the ratio $1 : \frac{1}{2} : \frac{1}{3} : ...$

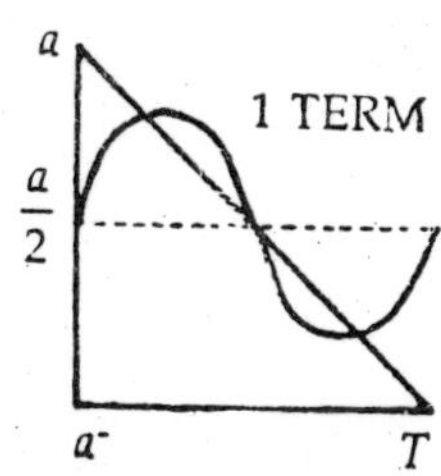

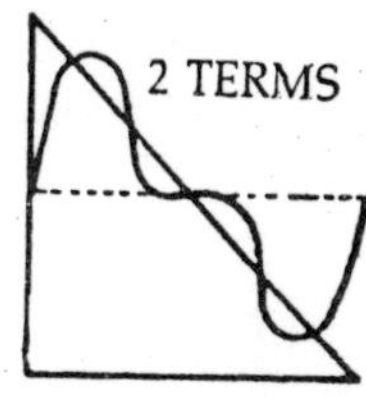

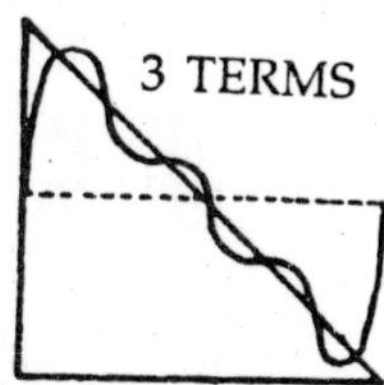

Fig.

The addition of successive terms of this series is indicated graphically in Fig. It is seen that greater the number of terms used, the closer the resemblances between the resultant curve and the curve under analysis.

Analysis of Saw-Tooth Function

The given function

$$y = a\left(1 - \frac{2t}{T}\right), \text{ for } 0 < t < T \qquad \text{...(i)}$$

represents a saw-tooth wave shown in Fig. It can be written as a Fouriers series

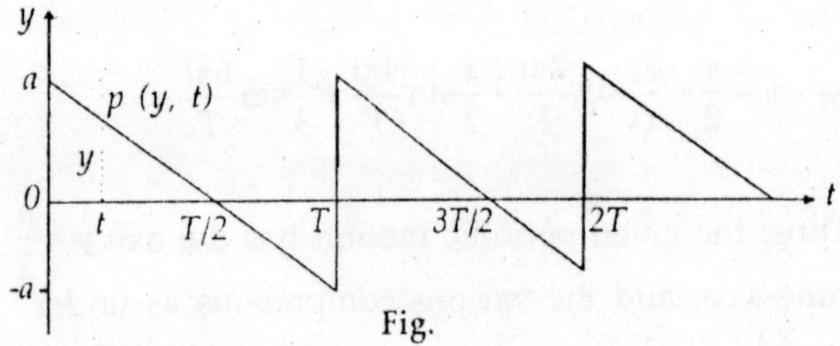

Fig.

$$y = A_0 + A_1 \cos\omega t + ... + A_r \cos r\omega t + ...$$
$$+ B_1 \sin\omega t + ... + B_r \sin r\omega t + ... \qquad \text{(ii)}$$

Let us evaluate the Fourier's coefficients A_0, A_r and B_r.

$$A_0 = \frac{1}{T}\int_0^T y \, dt$$

$$= \frac{1}{T}\int_0^T a\left(1 - \frac{2t}{T}\right) dt$$

$$= \frac{a}{T}\left[t - \frac{t^2}{T}\right]_0^T$$

$$= 0.$$

Proceeding exactly as in § 2.4, we find

$A_r = 0$

and $B_r = \dfrac{2a}{r\pi}$.

Thus, $B_1 = \dfrac{2a}{\pi},\ B_2 = \dfrac{2a}{2\pi},\ B_3 = \dfrac{2a}{3\pi}, ...$

Substituting these values in eq. (ii), we get

$$y = \frac{2a}{\pi}\left(\sin \omega t + \frac{1}{2}\sin 2\omega + \frac{1}{3}\sin 3\omega t + ...\right)$$

or $$y = \frac{2a}{\pi}\left(\sin\frac{2\pi t}{T} + \frac{1}{2}\sin\frac{4\pi t}{T} + \frac{1}{3}\sin\frac{6\pi t}{T} + ...\right).$$

The addition of successive terms of this series is shown graphically in Fig. It is seen that greater the number of terms used, the closer the resemblance between the resultant curve and the curve under analysis.

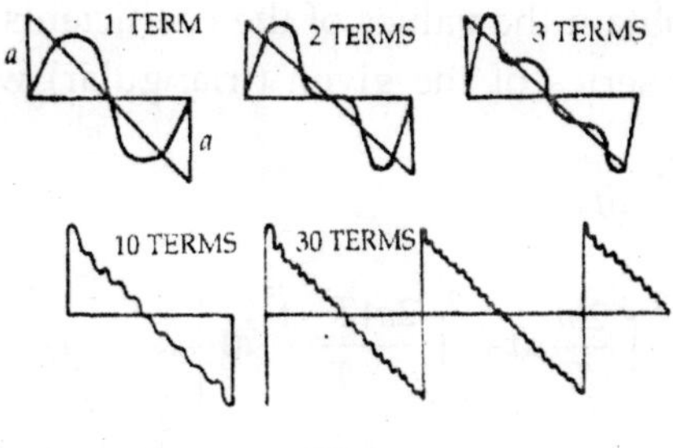

Fig.

(c) *Analysis of A Triangular Wave*

Let us consider triangular wave-form (Fig.),

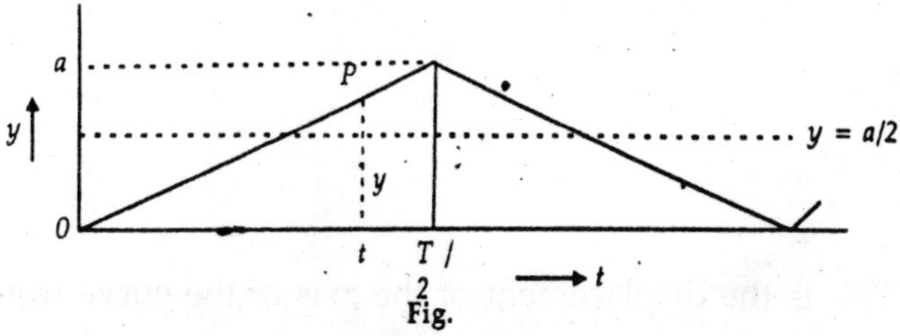

Fig.

and the curve repeats itself after $t = T$. The displacement at any time t may be written as

$$y = \frac{2at}{T} \quad \text{from } t = 0 \text{ to } t = T/2$$

and $$y = \frac{2a(T-t)}{T} \quad \text{from } t = T/2 \text{ to } t = T.$$

Now, the Fourier's series is

$$y = A_0 + A_1 \cos\omega t + \ldots + A_r \cos r\omega t + \ldots$$
$$+ B_1 \sin\omega t + \ldots + B_r \sin r\omega t + \ldots$$

where $A_0 = \frac{1}{T}\int_0^T y\,dt, A_r = \frac{2}{T}\int_0^T y\cos r\omega t\,dt$

and $B_r = \frac{2}{T}\int_0^T y\sin r\omega t\,dt.$

Let us obtain the values of the coefficients A_0, A_r and B_r of the Fourier series of the given (triangular) wave-form:

$$A_0 = \frac{1}{T}\int_0^T y\,dt$$

$$= \frac{1}{T}\left[\int_0^{T/2} \frac{2at}{T}dt + \int_{T/2}^{T} \frac{2a(T-t)}{T}dt\right]$$

$$= \frac{1}{T}\left[\frac{2a}{T}\left(\frac{t^2}{2}\right)_0^{T/2} + \frac{2a}{T}\left(Tt - \frac{t^2}{2}\right)_{T/2}^{T}\right]$$

$$= \frac{2a}{T^2}\left[\frac{T^2}{8} + \left(T^2 - \frac{T^2}{2}\right) - \left(\frac{T^2}{2} - \frac{T^2}{8}\right)\right]$$

$$= \frac{2a}{T^2}\left[\frac{T^2}{4}\right]$$

$$= \frac{a}{2}.$$

This is the displacement of the axis of the curve from the axis of the coordinate system. Now,

$$A_r = \frac{2}{T}\int_0^T y\cos r\omega t\,dt$$

$$= \frac{2}{T}\left[\int_0^{T/2} \frac{2at}{T}\cos r\omega t\;dt + \int_{T/2}^{T} \frac{2a(T-t)}{T}\cos r\omega t\,dt\right]$$

$$= \frac{4a}{T^2}\left[\int_0^{T/2} t\cos r\omega t\;dt + \int_{T/2}^{T} T\cos r\omega t\,dt - \int_{T/2}^{T} t\cos r\omega t\,dt\right]$$

On solving these integrals, we obtain

$$A_r = \frac{4a}{T^2}\left[\frac{2}{r^2\omega^2}\{(\cos \pi r)-1\}\right]$$

$$= \frac{4a}{T^2}\left[\frac{2}{r^2(2\pi/T)^2}\{(-1)^r-1\}\right] \quad [\because \cos \pi r = (-1)^r]$$

$$= \frac{2a}{\pi^2 r^2}\left[(-1)^r-1\right]$$

$= 0$, if r is even

$= -\frac{4a}{\pi^2 r^2}$, if r is odd.

Thus, only odd harmonics with cosine terms would appear in the Fourier series.

$$B_r = \frac{2}{T}\int_0^T y\sin r\omega t\,dt$$

$$= \frac{2}{T}\left[\int_0^{T/2} \frac{2at}{T}\sin r\omega t\;dt + \int_{T/2}^{T} \frac{2a(T-t)}{T}\sin r\omega t\,dt\right]$$

$$= \frac{4a}{T^2}\left[\int_0^{T/2} t\sin r\omega t\;dt + \int_{T/2}^{T} T\sin r\omega t\,dt - \int_{T/2}^{T} \sin r\omega t\,dt\right]$$

$= 0.$

Thus, all the sine terms in the Fourier series are zero.

Substituting the values of A_0, A_r and B_r in the Fourier series, we obtain

$$y = \frac{a}{2} - \frac{4a}{\pi^2}\cos\omega t - \frac{4a}{3^2\pi^2}\cos 3\omega t - \frac{4a}{5^2\pi^2}\cos 5\omega t \ldots$$

$$= \frac{a}{2} - \frac{4a}{\pi^2}\left(\cos\omega t + \frac{1}{3^2}\cos 3\omega t + \frac{1}{5^2}\cos 5\omega t + \ldots\right).$$

Thus, the given triangular vibration has the axis $y = \frac{a}{2}$ above the time-axis, and the various cosine components of amplitude coefficients,

$$-\frac{4a}{\pi^2}\left(1, \frac{1}{3^2}, \frac{1}{5^2}, \frac{1}{7^2}, \ldots\right)$$

and frequencies in the ratio 1 : 3 : 5 : 7....

The resultant obtained by adding three terms of this series is shown in Fig. As the series is rapidly convergent, the curve under analysis is closely represented by the sum of only a few terms of the series.

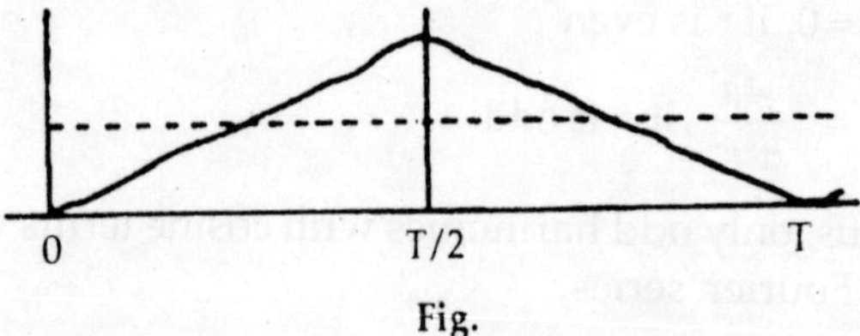

Fig.

(d) *Fourier's Analysis of the Output Wave From A Half-Wave Rectifier*

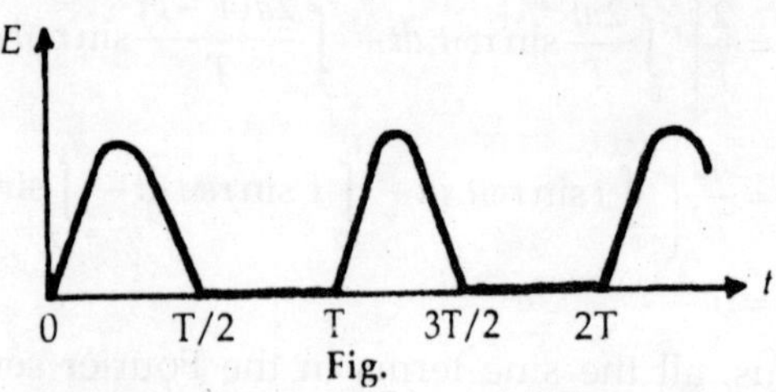

Fig.

A sinusoidal voltage $E = E_0 \sin\omega t$ is passed through a half-wave rectifier which removes the negative half-cycles of the wave. The output voltage wave is of the form shown in Fig. This may be expressed as

$$E(t) = \begin{cases} E_0 \sin \omega t & \text{from } t = 0 \text{ to } t = T/2 \\ 0 & \text{from } t = T/2 \text{ to } t/T \end{cases} \quad T = \frac{2\pi}{\omega}. \qquad \ldots\text{(i)}$$

Let us express it as a Fourier series:

$$E(t) = A_0 + A_1 \cos \omega t + \ldots + A_r \cos r\omega t + \ldots$$
$$+ B_1 \sin \omega t + \ldots + B_r \sin r\omega t + \ldots \qquad \ldots\text{(ii)}$$

Let us evaluate the Fourier coefficients:

$$A_0 = \frac{1}{T}\int_0^T E(t)\, dt$$

$$= \frac{1}{T}\int_0^{T/2} E_0 \sin \omega t\, dt$$

$$= \frac{1}{T}\int_0^{T/2} E_0 \sin \frac{2\pi t}{T} dt$$

$$= -\frac{E_0}{T}\frac{1}{2\pi/T}\left[\cos\frac{2\pi t}{T}\right]_0^{T/2}$$

$$= -\frac{E_0}{2\pi}[\cos\pi - \cos 0]$$

$$= \frac{E_0}{\pi}. \quad [\because \cos\pi = -1 \text{ and } \cos 0 = 1]$$

Again,

$$A_r = \frac{2}{T}\int_0^T E(t) \cos r\omega t\, dt$$

$$= \frac{2E_0}{T}\int_0^{T/2} \sin\omega t \cos r\omega t\, dt$$

$$= \frac{2E_0}{T}\int_0^{T/2} \sin\frac{2\pi t}{T}\cos\frac{2r\pi t}{T} dt$$

$$= \frac{2E_0}{T}\int_0^{T/2} \frac{1}{2}\left[\sin(1+r)\frac{2\pi t}{T} + \sin(1-r)\frac{2\pi t}{T}\right] dt$$

$$= -\frac{E_0}{T}\left[-\frac{\cos(1+r)2\pi t/T}{(1+r)2\pi/T} - \frac{\cos(1-r)2\pi t/T}{(1-r)2\pi/T}\right]_0^{T/2}$$

$$= -\frac{E_0}{T}\left[\frac{\cos(1+r)\pi - \cos 0}{(1+r)2\pi/T} + \frac{\cos(1-r)\pi - \cos 0}{(1-r)2\pi/T}\right]$$

$$= -\frac{E_0}{2\pi}\left[\frac{\cos(1+r)\pi - 1}{(1+r)} + \frac{\cos(1-r)\pi - 1}{(1-r)}\right].$$

When r is odd then this is equal to zero, because $\cos 0 = \cos 2\pi = \cos 4\pi = 1$. When r is even, then we have

$$A = \frac{E_0}{2\pi}\left[\frac{2}{1+r} + \frac{2}{1-r}\right] \quad [\because \cos\pi = \cos 3\pi = \cos 5\pi = -1]$$

$$= \frac{E_0}{\pi}\frac{2}{(1+r)(1-r)}$$

$$= -\frac{2E_0}{\pi}\frac{1}{(r-1)(r+1)}; \; r = 2, 4, 6, \ldots$$

$$= -\frac{2E_0}{\pi}\left[\frac{1}{(1)(3)}, \frac{1}{(3)(5)}, \frac{1}{(5)(7)}, \ldots\right].$$

Again,

$$B_r = \frac{2}{T}\int_0^T E(t)\sin r\omega t\, dt$$

$$= \frac{2E_0}{T}\int_0^{T/2} \sin\frac{2\pi t}{T}\sin\frac{2r\pi t}{T}\, dt$$

Proceeding as above, we can see that

$$B_r = \frac{E_0}{2} \text{ for } r = 1$$

and $B_r = 0$ for $r = 2, 3, 4, \ldots$.

Substituting these values of A_0, A_r, and B_r in eq. (ii), we get

$$E(t) = \frac{E_0}{\pi} + \frac{E_0}{2}\sin\omega t - \frac{2E_0}{\pi}\left[\frac{1}{1.3}\cos 2\omega t + \frac{1}{3.5}\cos 4\omega t + \frac{1}{5.7}\cos 6\omega t + \ldots\right].$$

(e) *Fourier Analysis of a Full Wave Rectifier Output*

The output voltage (or current) of a full-wave rectifier voltage is continuous, unidirectional but pulsating. It can be analysed by means of Fourier series.

Let $E = E_0 \sin \omega t$ be the applied input voltage to the rectifier. It r_p be the internal plate resistance of the diode, then the instantaneous output current through a load resistance R will be given by

$$i = \frac{E}{r_p + R} = \frac{E_0 \sin \omega t}{r_p + R}$$

$$= i_0 \sin \omega t,$$

where i_0 {$=E_0$ (r_p + R)} is the peak value of the current in the load.

Taking $f(\omega t) = i + i_0 \sin \omega t$, we can evaluate the coefficients A_0, A_r, B_r.

In the case of full-wave rectifier, i is positive for t = 0 to t = T/2, as well as for t = T/2 to t = T. That is,

$$i = i_0 \sin \omega t \quad \text{for } t = 0 \text{ to } t = T/2$$

and $\quad i = -i_0 \sin \omega t$ for t = T/2 to t = T,

because $\sin \omega t$ is negative in the range t = T/2 to t = T. Now, we have

$$i = f(t) = A_0 + \sum_{r=1}^{\infty} \left[A_r \cos(\omega t r) + B_r \sin(\omega r t) \right]$$

where

$$A_0 = \frac{2}{T} \left[\int_0^{T/2} i \sin \omega t \, dt - \int_{T/2}^{T} i \sin \omega t \, dt \right]$$

$$= \frac{\omega}{2\pi} \left[\int_0^{\pi/\omega} i_0 \sin \omega t \, dt - \int_0^{2\pi/\omega} i_0 \sin \omega t \, dt \right]$$

$$= \frac{\omega}{2\pi} \frac{i_0}{\omega} \left[\left| -\cos \omega t \right|_0^{\pi/\omega} - \left| \cos \omega t \right|_{\pi/\omega}^{2\pi/\omega} \right]$$

$$= \frac{i_0}{2\pi}\left[(-\cos\pi + \cos 0) - (-\cos 2\pi + \cos\pi)\right]$$

$$= \frac{2i_0}{2\pi}\left[(1+1) - (-1-1)\right]$$

$$= \frac{2i_0}{\pi}$$

$$A_r = \frac{2}{T}\left[\int_0^{T/2} i_0 \sin\omega t \cos r\omega t\, dt - \int_0^{T} i_0 \sin\omega t \cos r\omega t\, dt\right]$$

$$= \frac{\omega}{\pi}\frac{i_0}{2}\left[\int_0^{\pi/\omega}\{\sin(r+1)\omega t - \sin(r-1)\omega t\}dt\right.$$

$$\left. - \int_{\pi/\omega}^{2\pi/\omega}\{\sin(r+1)\omega t - \sin(r-1)\omega t\}dt\right]$$

$$= \frac{\omega}{\pi}\frac{i_0}{2\omega}\left[\left|-\frac{\cos(r+1)\omega t}{(r+1)\omega} + \frac{\cos(r-1)\omega t}{(r-1)\omega}\right|_0^{\pi/\omega}\right.$$

$$\left. - \left|-\frac{\cos(r+1)\omega t}{(r+1)\omega t} + \frac{\cos(r-1)\omega t}{(r-1)\omega}\right|_{\pi/\omega}^{2\pi/\omega}\right]$$

$$= \frac{\omega}{\pi}\frac{i_0}{2\omega}\left[\left\{-\frac{\cos(r+1)\pi}{r+1} + \frac{\cos(r-1)\pi}{r-1} + \frac{\cos 0}{r+1} - \frac{\cos 0}{r-1}\right\}\right.$$

$$\left. - \left\{-\frac{\cos(r+1)2\pi}{r+1} + \frac{\cos(r-1)2\pi}{r-1} + \frac{\cos(r+1)\pi}{r+1} - \frac{\cos(r-1)\pi}{r-1}\right\}\right]$$

Now, when r is odd

$$A_r = \frac{i_0}{2\pi}\left[\left\{-\frac{1}{r+1} + \frac{1}{r-1} + \frac{1}{r+1} - \frac{1}{r-1}\right\}\right.$$

$$\left. - \left\{-\frac{1}{r+1} + \frac{1}{r-1} + \frac{1}{r+1} - \frac{1}{r-1}\right\}\right]$$

$$= 0$$

When r is even:

$$A_r = \frac{i_0}{2\pi}\left[\left\{-\frac{1}{r+1} - \frac{1}{r-1} + \frac{1}{r+1} - \frac{1}{r-1}\right\}\right.$$

$$-\left\{-\frac{1}{r+1}+\frac{1}{r-1}-\frac{1}{r+1}+\frac{1}{r-1}\right\}\Bigg]$$

$$=\frac{4i_0}{2\pi}\left[\frac{1}{r+1}-\frac{1}{r-1}\right]$$

$$=-\frac{4i_0}{\pi\left(r^2-1\right)}$$

Thus, the values of the coefficients A's are (r = 2, 4, 6,...)

$$-\frac{4i_0}{3\pi},-\frac{4i_0}{15\pi},-\frac{4i_0}{35\pi},\ldots$$

$$B_r=\frac{2}{T}\left[\int_0^{T/2} i_0 \sin\omega t \sin r\omega t\, dt-\int_{T/2}^{T} i_0 \sin\omega t \sin r\omega t\, dt\right]$$

$= 0$ (it can be worked out exactly as above).

Thus, the output current i can be expressed in the form of a Fourier series as

$$i=\frac{2i_0}{\pi}-\frac{4i_0}{3\pi}\cos 2\omega t-\frac{4i_0}{15\pi}\cos 4\omega t-\frac{4i_0}{35\pi}\cos 6\omega t\ldots$$

Thus, it consists of a d.c. component $\frac{2i_0}{\pi}$ and a series of a.c. components (even harmonics only) or ripples.

Operative Areas

Particular Areas for Operation

The three quantities which have been specially studied in this chapter are $\nabla\phi$, dif f and curl f. For these quantities certain important points are to be noted: i) grad is associated with a *scalar point function* f and grad f is a *vector*; ii) Divergence is associated with vector point function f and divergence f is a *scalar* and iii) Curl is associated with vector function f and curl f is a *vector.*

Since grad ϕ and curl f are vector point functions and as such we can find their divergence as well as curl. Also div f is a scalar point function we can find its grad we may thus form the following functions:

curl grad $\phi=\nabla\times(\nabla\phi)$

div curl f $=\nabla.(\nabla\times f)$

div grad $\phi=\nabla.(\nabla\phi)$

curl curl $f=\nabla\times(\nabla\times f)$

grad div $f=\nabla(\nabla.f)$

These are called second order differential functions

Properties of Second Order Differential Operators

Property 1: curl (grad ϕ) = $\nabla\times(\nabla\phi)=0$.

We have

$$\nabla\times(\nabla\phi) = \nabla\times\left(i\frac{\partial\phi}{\partial x}+j\frac{\partial\phi}{\partial y}+k\frac{\partial\phi}{\partial z}\right)$$

$$= \begin{vmatrix} i & j & k \\ \frac{\partial}{\partial x} & \frac{\partial}{\partial y} & \frac{\partial}{\partial z} \\ \frac{\partial\phi}{\partial x} & \frac{\partial\phi}{\partial y} & \frac{\partial\phi}{\partial z} \end{vmatrix}$$

$$= i\left(\frac{\partial^2\phi}{\partial y\partial z}-\frac{\partial^2\phi}{\partial z\partial y}\right)+j\left(\frac{\partial^2\phi}{\partial z\partial x}-\frac{\partial^2\phi}{\partial x\partial z}\right)+k\left(\frac{\partial^2\phi}{\partial x\partial y}-\frac{\partial^2\phi}{\partial y\partial x}\right)$$

$$= 0,$$

as $\frac{\partial^2 y}{\partial y\partial z}=\frac{\partial^2\phi}{\partial z\partial y}$, etc.

Hence, $\nabla\times(\nabla\phi)=0$.

Property 2: *div* (curl f) = $\nabla\times(\nabla f)=0$.

Let $f = if_1 + jf_2 + kf_3$

$$\therefore \quad \nabla\times f = \begin{vmatrix} i & j & k \\ \frac{\partial}{\partial x} & \frac{\partial}{\partial y} & \frac{\partial}{\partial z} \\ f_1 & f_2 & f_3 \end{vmatrix}$$

$$= i\left(\frac{\partial f_3}{\partial y}-\frac{\partial f_2}{\partial z}\right)+j\left(\frac{\partial f_1}{\partial z}-\frac{\partial f_3}{\partial x}\right)+k\left(\frac{\partial f_2}{\partial x}-\frac{\partial f_1}{\partial y}\right)$$

Therefore

$$\nabla.(\nabla\times f)=\frac{\partial}{\partial x}\left(\frac{\partial f_3}{\partial y}-\frac{\partial f_2}{\partial z}\right)+\frac{\partial}{\partial y}\left(\frac{\partial f_1}{\partial z}-\frac{\partial f_3}{\partial x}\right)+\frac{\partial}{\partial z}\left(\frac{\partial f_2}{\partial x}-\frac{\partial f_1}{\partial y}\right)$$

$$=\frac{\partial^2 f_3}{\partial x\partial y}-\frac{\partial^2 f_2}{\partial x\partial z}+\frac{\partial^2 f_1}{\partial y\partial x}-\frac{\partial^2 f_3}{\partial y\partial x}+\frac{\partial^2 f_2}{\partial y\partial x}-\frac{\partial^2 f_1}{\partial z\partial y}$$

$$=0$$

Hence, $\nabla.(\nabla\times f)=0$

Property 3: div (grad ϕ) = $\nabla.(\nabla\phi)=\frac{\partial^2\phi}{\partial x^2}+\frac{\partial^2\phi}{\partial y^2}+\frac{\partial^2\phi}{\partial z^2}$

We have

$$\nabla.(\nabla\phi)=\left(i\frac{\partial}{\partial x}+j\frac{\partial}{\partial y}+k\frac{\partial}{\partial z}\right).\left(i\frac{\partial\phi}{\partial x}+j\frac{\partial\phi}{\partial y}+k\frac{\partial\phi}{\partial z}\right)$$

$$=\frac{\partial^2\phi}{\partial x^2}+\frac{\partial^2\phi}{\partial y^2}+\frac{\partial^2\phi}{\partial z^2}=\nabla^2\phi$$

$$\nabla.(\nabla\phi)=\nabla^2\phi$$

Property 4: curl (curl f) = grad (div f) – $\nabla^2\phi$

i.e., $\nabla\times(\nabla\times f)=\nabla(\nabla.f)-\nabla^2 f$

Let $f=if_1+jf_2+kf_3$

$$\nabla\times f=i\left(\frac{\partial f_3}{\partial y}-\frac{\partial f_2}{\partial z}\right)+j\left(\frac{\partial f_1}{\partial z}-\frac{\partial f_3}{\partial x}\right)+k\left(\frac{\partial f_2}{\partial x}-\frac{\partial f_1}{\partial y}\right)$$

$$\nabla\times(\nabla\times f)=\begin{vmatrix} i & j & k \\ \partial/\partial x & \partial/\partial y & \partial/\partial z \\ \frac{\partial f_3}{\partial y}-\frac{\partial f_1}{\partial z} & \frac{\partial f_1}{\partial z}-\frac{\partial f_3}{\partial x} & \frac{\partial f_2}{\partial z}-\frac{\partial f_1}{\partial y}\end{vmatrix}$$

$$= \mathrm{i}\left\{\frac{\partial}{\partial y}\left(\frac{\partial f_2}{\partial x}-\frac{\partial f_1}{\partial y}\right)-\frac{\partial}{\partial z}\left(\frac{\partial f_1}{\partial z}-\frac{\partial f_3}{\partial x}\right)\right\}$$

$$+\mathrm{j}\left\{\frac{\partial}{\partial z}\left(\frac{\partial f_3}{\partial y}-\frac{\partial f_2}{\partial z}\right)-\frac{\partial}{\partial x}\left(\frac{\partial f_2}{\partial x}-\frac{\partial f_1}{\partial y}\right)\right\}$$

$$+\mathrm{k}\left\{\frac{\partial}{\partial x}\left(\frac{\partial f_1}{\partial z}-\frac{\partial f_3}{\partial x}\right)-\frac{\partial}{\partial y}\left(\frac{\partial f_3}{\partial y}-\frac{\partial f_2}{\partial z}\right)\right\}$$

$$= \mathrm{i}\left\{\frac{\partial^2 f_2}{\partial y\partial x}-\frac{\partial^2 f_1}{\partial y^2}-\frac{\partial^2 f_1}{\partial z^2}+\frac{\partial^2 f_2}{\partial y\partial x}\right\}$$

$$+\mathrm{j}\left\{\frac{\partial^2 f_3}{\partial z\partial y}-\frac{\partial^2 f_2}{\partial z^2}-\frac{\partial^2 f_2}{\partial x^2}+\frac{\partial^2 f_1}{\partial x\partial y}\right\}$$

$$+\mathrm{k}\left\{\frac{\partial^2 f_1}{\partial x\partial z}-\frac{\partial^2 f_3}{\partial x^2}-\frac{\partial^2 f_3}{\partial y^2}+\frac{\partial^2 f_2}{\partial y\partial z}\right\}$$

$$= \mathrm{i}\frac{\partial}{\partial x}\left(\frac{\partial f_1}{\partial x}+\frac{\partial f_2}{\partial y}+\frac{\partial f_3}{\partial z}\right)$$

$$+\mathrm{j}\frac{\partial}{\partial y}\left(\frac{\partial f_1}{\partial x}+\frac{\partial f_2}{\partial y}+\frac{\partial f_3}{\partial z}\right)$$

$$+\mathrm{k}\frac{\partial}{\partial z}\left(\frac{\partial f_1}{\partial x}+\frac{\partial f_2}{\partial y}+\frac{\partial f_3}{\partial z}\right)$$

$$\left(i\nabla^2 f_1+i\nabla f_2+k\nabla^2 f_3\right)$$

$$= \nabla(\nabla.\mathrm{f})-\nabla^2\left(\mathrm{i}f_1+\mathrm{i}f_2+\mathrm{k}f_3\right)$$

$$= \nabla(\nabla.\mathrm{f})-\nabla^2\mathrm{f}$$

$$\nabla\times(\nabla.\mathrm{f}) = \nabla\times(\nabla.\mathrm{f})-\nabla^2\mathrm{f}$$

ILLUSTRATIVE EXAMPLES

1. If $\phi = x^3 + y^3 + z^3 - 3xyz$, find div grad ϕ and curl grad ϕ.

Solution: We have

$$\text{grad}\,\phi = i\frac{\partial\phi}{\partial x} + j\frac{\partial\phi}{\partial y} + k\frac{\partial\phi}{\partial z}$$

$$= i\left(3x^2 - 3yz\right) + j\left(3y^2 - 3zx\right) + k\left(3z^2 - 3xy\right)$$

$$\text{div grad}\,\phi = \frac{\partial}{\partial x}\left(3x^2 - 3yz\right) + \frac{\partial}{\partial y}\left(3y^2 - 3zx\right) + \frac{\partial}{\partial z}\left(3z^2 - 3xy\right)$$

$$= 6x + 6y + 6z = 6\left(x + y + z\right)$$

Again
$$\text{curl}\ \nabla\phi = \begin{vmatrix} i & j & k \\ \dfrac{\partial}{\partial x} & \dfrac{\partial}{\partial y} & \dfrac{\partial}{\partial z} \\ 3x^2 - 3yz & 3y^2 - 3zx & 3z^2 - 3xy \end{vmatrix}$$

$$= i\left\{\frac{\partial}{\partial y}\left(3z^2 - 3xy\right) - \frac{\partial}{\partial z}\left(3y^2 - 3zx\right)\right\}$$

$$+ j\left\{\frac{\partial}{\partial y}\left(3x^2 - 3yz\right) - \frac{\partial}{\partial x}\left(3z^2 - 3xy\right)\right\}$$

$$+ k\left\{\frac{\partial}{\partial x}\left(3y^2 - 3zx\right) - \frac{\partial}{\partial y}\left(3x^2 - 3yz\right)\right\}$$

$$= i(-3x + 3x) + j(-3y + 3y) + k(-3z + 3z)$$

$$= 0.$$

2. Prove that div $\nabla r^n = \nabla^2 r^n = n(n+1)r^{n-2}$

Solution: Let r = xi + yj + zk

$$\therefore \qquad r = \sqrt{\left(x^2 + y^2 + z^2\right)}$$

$$\therefore \quad \frac{\partial r}{\partial x} = \frac{2x}{2\sqrt{(x^2+y^2+z^2)}}$$

$$= \frac{x}{r}, \frac{\partial r}{\partial y} = \frac{y}{r} \text{ and } \frac{\partial r}{\partial y} = \frac{z}{r} \quad \ldots\text{(i)}$$

Let $\phi = r^n$, then

$$\frac{\partial \phi}{\partial x} = nr^{n-1}\frac{\partial r}{\partial x} = nr^{n-1}.\frac{x}{r} = nxr^{n-2} \quad \ldots\text{(ii)}$$

$$\frac{\partial^2 \phi}{\partial x^2} = n\left[1.r^{n-2} + x(n-2)r^{n-3}.\frac{x}{r}\right]$$

$$= nr^{n-2}\left[1 + \frac{(n-2)}{r^2}x^2\right]$$

Similarly, $\frac{\partial^2 \phi}{\partial y^2} = nr^{n-2}\left[1 + \frac{(n-2)}{r^2}y^2\right]$

and $\frac{\partial^2 \phi}{\partial y^2} = nr^{n-2}\left[1 + \frac{(n-2)}{r^2}z^2\right]$

$$\therefore \quad \text{div grad } r^n = \nabla^2\phi = \frac{\partial^2\phi}{\partial x^2} + \frac{\partial^2\phi}{\partial y^2} + \frac{\partial^2\phi}{\partial z^2}$$

$$= nr^{n-2}\left[1 + \frac{(n-2)}{r^2}x^2\right]$$

$$+ nr^{n-2}\left[1 + \frac{(n-2)}{r^2}y^2\right]$$

$$+ nr^{n-2}\left[1 + \frac{(n-2)}{r^2}z^2\right]$$

$$= nr^{n-2}\left[3 + \frac{(n-2)}{r^2}(x^2+y^2+z^2)\right]$$

$$= nr^{n-2}\left[3+\frac{(n-2)}{r^2}.r^2\right]$$

$$= nr^{n-2}(3+n-2)$$

$$= n(n+1)r^{n-2}.$$

3. *Prove that* $\nabla^2(r^n r) = n(n+3)r^{n-2}r$.

Solution: $\text{L.H.S.} = \frac{\partial^2}{\partial x^2}(r^n r)+\frac{\partial^2}{\partial y^2}(r^n r)+\frac{\partial^2}{\partial z^2(r^n r)}$

$$= \frac{\partial}{\partial x}\left(nr^{n-1}\frac{\partial r}{\partial x}r+r^n\frac{\partial}{\partial x}\right)+\frac{\partial}{\partial y}\left(nr^{n-1}\frac{\partial r}{\partial y}r+r^n\frac{\partial r}{\partial y}\right)$$

$$+\frac{\partial}{\partial z}\left(nr^{n-1}\frac{\partial r}{\partial z}r+r^n\frac{\partial r}{\partial z}\right)$$

$$= \frac{\partial}{\partial x}\left(nr^{n-1}\frac{x}{r}r+r^n i\right)+\frac{\partial}{\partial y}\left(nr^{n-1}\frac{y}{r}.r+r^n j\right)+\frac{\partial}{\partial z}\left(nr^{n-1}\frac{z}{r}.r+r^n k\right)$$

$$= \frac{\partial}{\partial x}(nr^{n-2}xr+r^n i)+\frac{\partial}{\partial y}(nr^{n-2}y.r+r^n j)+\frac{\partial}{\partial z}(nr^{n-2}z.r+r^n k)$$

$$= \left[n\left\{1.r^{n.2}r+xr(n-2)r^{n-3}\frac{x}{r}+r^{n-2}xi\right\}+nr^{n-1}\frac{x}{r}i\right]$$

$$+\left[n\left\{1.r^{n-2}r+yr(n-2)r^{n-3}.\frac{y}{r}+r^{n-2}yj\right\}+nr^{n-1}\frac{y}{r}j\right]$$

$$+\left[n\left\{1.r^{n-2}r+zr(n-2)r^{n-3}.\frac{z}{r}+r^{n-2}zk\right\}+nr^{n-1}\frac{z}{r}k\right]$$

$$= n\left[3r^{n-2}r+(n-2)r^{n-4}r(x^2+y^2+z^2)+r^{n-2}(xi+yj+zk)\right]$$

$$+nr^{n-2}(xi+yj+zk)$$

$$= n\left[3r^{n-2}r+(n-2)r^{n-2}r+r^{n-2}r\right]+nr^{n-2}r$$

$$= nr^{n-2}r[3+n-2+1+1]$$

$$= n(n+3)r^{n-2}r.$$

EXERCISE

1. Verify that curl grad $\phi = 0$ and div $\phi A = \nabla\phi.A + \phi$ div A, given that $\phi = x^3 + y^3 + z^3 + 3xyz$ and $A = x^2i + y^2j + z^2k.$
2. Prove that curl grad r^n = 0.
3. If f = x^2yi + xzj + $2yz$k, Prove that div (curl f) = 0.
4. Prove that div grad $\left(\frac{1}{r}\right) = 0.$
5. If $\mathrm{f} = \frac{\mathrm{r}}{\mathrm{r}}$, then show that grad div $\mathrm{f} = -\frac{2\mathrm{r}}{r^3}.$
6. Show that $\nabla^2 f(r) = f''(r) + \frac{2}{r}f'(r)$
7. Prove that $\nabla^2(r\,\mathrm{r}) = 4\hat{\mathrm{r}} = \frac{4\hat{\mathrm{r}}}{r}.$
8. Prove that curl $\{f(r)\mathrm{r}\} = 0.$

Vector Identities: If u and v be two scalar functions and a and b be two vector functions, then we can have the products u v and a b both scalar. So we shall find grad (u v) and grad (a b). Similarly products ua and a × b are vectors, so we can find both their divergence as well as curl, *i.e.*,

div (u a), div (a × b) and curl (u a), curl (a × b)

These results are known as *vector identities* and we shall find these one by one.

(I) grad (u v) = u grad v + u grad v

We have

$$\nabla(uv) = \Sigma \mathrm{i}\frac{\partial}{\partial x}(uv)$$

$$= \Sigma \mathrm{i}\left(\mathrm{u}\frac{\partial v}{\partial x} + v\frac{\partial u}{\partial x}\right)$$

$$= u\left(\mathrm{i}\frac{\partial v}{\partial x} + \mathrm{j}\frac{\partial v}{\partial y} + \mathrm{k}\frac{\partial v}{\partial z}\right) + v\left(\mathrm{i}\frac{\partial u}{\partial x} + \mathrm{j}\frac{\partial u}{\partial y} + \mathrm{k}\frac{\partial u}{\partial z}\right)$$

$$= u\nabla v + v\nabla u$$

i.e., $\nabla(uv) = u\nabla v + v\nabla u$

(II) grad (a. b) = a × curl b + b × curl a + (a. ∇) b + (b. ∇) a

We have $\nabla(\mathrm{a.b}) = \Sigma \mathrm{i}\frac{\partial}{\partial x}(\mathrm{a} \,.\, \mathrm{b})$

$$= \Sigma \mathrm{i}\left(\mathrm{a}.\frac{\partial \mathrm{b}}{\partial x} + \mathrm{b}.\frac{\partial \mathrm{a}}{\partial x}\right)$$

$$= \Sigma \mathrm{i}\left(\mathrm{a}.\frac{\partial \mathrm{b}}{\partial x}\right) + \Sigma \mathrm{i}\left(\mathrm{b}.\frac{\partial \mathrm{a}}{\partial x}\right) \qquad \ldots(\mathrm{i})$$

Now, $\mathrm{a} \times \left(\mathrm{i} \times \frac{\partial \mathrm{b}}{\partial x}\right) = \left(\mathrm{a}.\frac{\partial \mathrm{b}}{\partial x}\right)\mathrm{i} - (\mathrm{a} \,.\, \mathrm{i})\frac{\partial \mathrm{b}}{\partial x}$

or $\left(\mathrm{a}.\frac{\partial \mathrm{b}}{\partial x}\right)\mathrm{i} = \mathrm{a} \times \left(\mathrm{i} \times \frac{\partial \mathrm{b}}{\partial x}\right) + (\mathrm{a} \,.\, \mathrm{i})\frac{\partial \mathrm{b}}{\partial x}$

$\therefore$ $\Sigma\left(\mathrm{a}\,.\frac{\partial \mathrm{b}}{\partial x}\right)\mathrm{i} = \mathrm{a} \times \Sigma\left(i \times \frac{\partial \mathrm{b}}{\partial x}\right) + \Sigma(\mathrm{a} \,.\, \mathrm{i})\frac{\partial \mathrm{b}}{\partial x}$

$$= \mathrm{a} \times \mathrm{curl}\ \mathrm{b} + (\mathrm{a}.\ \nabla)\ \mathrm{b} \qquad \ldots(\mathrm{ii})$$

Similarly,

$$\Sigma\left(b\,.\frac{\partial a}{\partial x}\right)i \;=\; b \times \text{curl}\, a + (b\,.\,\nabla)\, b \qquad \text{...(iii)}$$

Hence, from (i), (ii), and (iii) we obtain

grad $(a.\ b) = a \times \text{curl}\ b + \text{b} \times \text{curl}\ a + (a.\nabla)b + (b.\nabla)a$

(III) div (u a) = u div a + a. ∇ u.

We have

$$\text{div }(u\ a) = \nabla.(ua) = i.\frac{\partial}{\partial x}(ua) + j.\frac{\partial}{\partial y}(ua) + k.\frac{\partial}{\partial z}(ua)$$

$$= i.\left(a\frac{\partial u}{\partial x} + u\frac{\partial a}{\partial x}\right) + j.\left(a\frac{\partial a}{\partial y} + u\frac{\partial a}{\partial y}\right) + k.\left(a\frac{\partial u}{\partial x} + u\frac{\partial a}{\partial z}\right)$$

$$= i.\left(a\frac{\partial u}{\partial x}\right) + j.\left(a\frac{\partial u}{\partial y}\right) + k.\left(a\frac{\partial u}{\partial z}\right) + u\left(i\frac{\partial a}{\partial x} + j.\frac{\partial a}{\partial y} + k.\frac{\partial a}{\partial z}\right)$$

$$= a.\left(i\frac{\partial u}{\partial x} + j.\frac{\partial u}{\partial y} + k.\frac{\partial u}{\partial z}\right) + u\left(i\frac{\partial a}{\partial x} + j.\frac{\partial a}{\partial y} + k.\frac{\partial a}{\partial z}\right)$$

$= a.$ grad $u + u$ div a.

(IV) div (a × b) = b. curl a – a curl b

We have

$$\text{div }(a \times b) = i \times \frac{\partial}{\partial x}(a \times b) + j \times \frac{\partial}{\partial y}(a \times b) + k\frac{\partial}{\partial z}(a \times b)$$

$$= \Sigma i \times \left(\frac{\partial a}{\partial x} \times b + a \times \frac{\partial b}{\partial x}\right)$$

$$= \sum i.\frac{\partial a}{\partial x} \times b + a \times \frac{\partial b}{\partial x}$$

$$= \left(\sum i.\times\frac{\partial a}{\partial x}\right).b - \left(\sum i.\frac{\partial b}{\partial x}\right).a$$

$= b.$ curl $a - a.$ curl b

(v) curl (u a) = grad u) × a × u curl a

We have

$$\text{curl (u a)} = i\times\frac{\partial}{\partial x}(ua)+j\times\frac{\partial}{\partial y}(ua)+k\times\frac{\partial}{\partial z}(ua)$$

$$= \sum i\times\left(\frac{\partial u}{\partial x}a+u\frac{\partial a}{\partial x}\right)$$

$$= \sum i\times\left(\frac{\partial u}{\partial x}a\right)+\sum i\times u\frac{\partial a}{\partial x}$$

$$= \sum\left(i-\frac{\partial u}{\partial x}\right)\times a+\left(\sum i\times\frac{\partial a}{\partial x}\right)u$$

= (grad u) × a + u curl a.

(VI) curl (a × b) = a div b – b div a + (b. ∇) a - (a. ∇) b

We have

$$\text{Curl (a×b)} = i\times\frac{\partial}{\partial x}(a\times b)+j\times\frac{\partial}{\partial y}(a\times b)\times\frac{\partial}{\partial z}(a\times b)$$

$$= \Sigma i\times\left(a\times\frac{\partial b}{\partial x}+\frac{\partial a}{\partial x}\times b\right)$$

$$= \Sigma i\times\left(a\times\frac{\partial b}{\partial x}\right)+\Sigma i+\left(\frac{\partial a}{\partial x}\times b\right)$$

$$= \Sigma\left(i\,.\frac{\partial b}{\partial x}\right)a-\Sigma(i\,.\,a)\frac{\partial b}{\partial x}+\Sigma(i\,.\,b)\frac{\partial a}{\partial x}-\Sigma\left(i.\frac{\partial a}{\partial x}\right)b$$

as a × (b × c) = (a. c) b – (a. b) c

$$= \left(\Sigma i.\frac{\partial b}{\partial x}\right)a-\left(\Sigma i.\frac{\partial a}{\partial x}\right)b+\Sigma(i\,.\,b)\frac{\partial a}{\partial x}-\Sigma\left(i.\frac{\partial a}{\partial x}\right)b$$

= a div b – b a + (b . ∇) a - (a . v)b

ILLUSTRATIVE EXAMPLES

1. If a be a constant vector find grad (a. f). div (a × f) and curl (a × f).

Solution: We have

$$\text{grad (a. f)} = \text{a} \times \text{curl f} + \text{f} \times \text{curl a} + (\text{a} \,.\, \nabla)\,\text{f} + (\text{f} \,.\, \nabla)\text{a},$$

[by Identity II]

$$= \text{a} \times \text{curl f} + (\text{a} \,.\, \nabla)\,\text{f}$$

as a is constant vector, hence curl a = 0

and $(\text{f} \,.\, \nabla)\,\text{a} = 0$

$$\text{div (a} \times \text{f)} = (\text{curl a}) \,.\, \text{f} - (\text{curl f}) \,.\, \text{a}$$ [by Identity IV]

$$= -\,(\text{curl f}) \,.\, \text{a as curl a} = 0$$

$$\text{curl (a} \times \text{f)} = \text{a div f} - \text{f div a} + (\text{f} \,.\, \nabla)\,\text{a} - (\text{a} \,.\, \nabla)\,\text{f}$$

[by Identity VI]

$$= \text{a div f} - (\text{a} \,.\, \nabla)\,\text{f}$$

as div a = 0 and $(\text{f} \,.\, \nabla)\,\text{a} = 0$

2. If $f = \psi$ gradϕ, show that f. curl f = 0

Solution: We have

$$\text{curl f} = \text{curl}\,(\psi \text{grad}\phi)$$

$$= (\text{grad}\psi) \times (\text{grad}\phi) + \psi \text{curl}(\text{grad}\phi)$$

$$\left[\text{as curl}\, u\,\text{a} = (\nabla n) \times \text{a} + u\,\text{curl a}\right]$$

$$= (\text{grad}\psi) \times (\text{grad}\phi) \text{ as curl } (\nabla\phi) = 0$$

$$\therefore \quad \text{f . curl f} = \psi(\nabla\phi).\{(\nabla\psi) \times (\nabla\phi)\} = 0$$

as scalar triple product in which two vectors are equal is zero.

3. *If $\phi = P(x, y, z)i + Q(x, y, z)j + R(x, y, z)k$ verify that curl $\phi = i \times \phi_x + j \times \phi_y k \times \phi_z$.*

Solution: We have

$$\text{curl } f = \begin{vmatrix} i & j & k \\ \frac{\partial}{\partial x} & \frac{\partial}{\partial y} & \frac{\partial}{\partial z} \\ P & Q & R \end{vmatrix}$$

$$= i\left(\frac{\partial R}{\partial y} - \frac{\partial Q}{\partial z}\right) + j\left(\frac{\partial P}{\partial z} - \frac{\partial R}{\partial x}\right) + k\left(\frac{\partial Q}{\partial x} - \frac{\partial P}{\partial y}\right) \quad \ldots(i)$$

Also $\quad i \times \phi_y = \left(i\frac{\partial P}{\partial x} + j\frac{\partial Q}{\partial x} + k\frac{\partial R}{\partial x}\right)$

$$= k\frac{\partial Q}{\partial x} - j\frac{\partial Q}{\partial z}$$

Similarly,

$$j \times \phi_y = i\frac{\partial R}{\partial y} - k\frac{\partial Q}{\partial z}$$

and $k \times \phi_z = j\frac{\partial P}{\partial z} - i\frac{\partial Q}{\partial z}$

Hence, $\quad i \times \phi_x + j \times \phi_y + k \times \phi_z$

$$= i\left(\frac{\partial R}{\partial y} - \frac{\partial Q}{\partial z}\right) + j\left(\frac{\partial P}{\partial z} - \frac{\partial R}{\partial x}\right) + k\left(\frac{\partial Q}{\partial x} - \frac{\partial P}{\partial y}\right)$$

$= \text{curl}\,\phi$, From (i)

4. *If*

$$F = \left(y\frac{\partial f}{\partial z} - z\frac{\partial f}{\partial y}\right)i + \left(z\frac{\partial f}{\partial x} - x\frac{\partial f}{\partial z}\right)j + \left(x\frac{\partial f}{\partial y} - y\frac{\partial f}{\partial x}\right)k$$

then prove that (i) $F = r \times \nabla f$, ***(ii)*** $F.r = 0$ and ***(iii)*** $F.grad\ f = 0$

Solution: We have

$$r = ix \times jy + kz$$

and $$\nabla f = i\frac{\partial f}{\partial x} + j\frac{\partial f}{\partial y} + z\frac{\partial f}{\partial z}$$

(i) $$r \times \nabla f = \begin{vmatrix} i & j & k \\ x & y & z \\ \frac{\partial f}{\partial x} & \frac{\partial f}{\partial y} & \frac{\partial f}{\partial z} \end{vmatrix}$$

$$= i\left(y\frac{\partial f}{\partial z} - z\frac{\partial f}{\partial x}\right) + j\left(z\frac{\partial f}{\partial x} - x\frac{\partial f}{\partial z}\right) + k\left(x\frac{\partial f}{\partial y} - y\frac{\partial f}{\partial x}\right)$$

$$= F$$

(ii) $$F.r. = \left[\left(y\frac{\partial f}{\partial z} - z\frac{\partial f}{\partial x}\right)i + \left(z\frac{\partial f}{\partial x} - x\frac{\partial f}{\partial z}\right)j\right.$$

$$\left. + \left(x\frac{\partial f}{\partial y} - y\frac{\partial f}{\partial x}\right)k\right].(ix + jy + kz.)$$

$$= \left(xy\frac{\partial f}{\partial z} - xz\frac{\partial f}{\partial y} + yz\frac{\partial f}{\partial x} - xy\frac{\partial f}{\partial z} + xz\frac{\partial f}{\partial y} - yz\frac{\partial f}{\partial x}\right)$$

$$= 0$$

(iii) $$F.\text{grad} f = \left[\left(y\frac{\partial f}{\partial z} - z\frac{\partial f}{\partial x}\right)i + \left(z\frac{\partial f}{\partial x} - x\frac{\partial f}{\partial z}\right)j\right.$$

$$\left. + \left(x\frac{\partial f}{\partial y} - y\frac{\partial f}{\partial x}\right)k\right].\left(i\frac{\partial f}{\partial x} + j\frac{\partial f}{\partial y} + k\frac{\partial f}{\partial z}\right)$$

$$= \frac{\partial f}{\partial x}.\frac{\partial f}{\partial z} - z\frac{\partial f}{\partial x}.\frac{\partial f}{\partial y} + z\frac{\partial f}{\partial x}.\frac{\partial f}{\partial y}$$

$$-x\frac{\partial f}{\partial y}.\frac{\partial f}{\partial z}+x\frac{\partial f}{\partial y}.\frac{\partial f}{\partial z}-y\frac{\partial f}{\partial x}.\frac{\partial f}{\partial z}$$

EXERCISE

1. If f and g are irretational, show that f × g is solenoidal.
2. Find $\nabla.(A \times r)$, if curl A = 0.
3. Prove that $\nabla.\left[r\nabla\left(\frac{1}{r^3}\right)\right]=\frac{3}{r^4}$
4. Prove that $\nabla^2\left[\nabla\left(\frac{r}{r^2}\right)\right]=2r^{-4}$
5. Prove that $\nabla^2(uv)=u\nabla^2u+2\nabla v.\nabla v+\nabla u$.
6. Prove that $\nabla\left(a.\frac{r}{r^n}\right)=\frac{a}{r^n}-(a.r)r$
7. Prove that div $\frac{a \times r}{r^3}=0$ and curl $\frac{a \times r}{r^3}=-\frac{a}{r^3}+\frac{3r}{r^5}(a.r)$.
8. Prove that $\nabla.\left(a \times \frac{r}{r^n}\right)=0$
9. Show that curl (a . r) a = 0
10. Prove that curl $(\psi\nabla\psi)=\nabla\psi\times\nabla\psi=-\text{curl}(\psi\nabla\psi)$.

Integrals: Line, Volume and Surface

Line Integrals: Let r = f (t) represents, a continuously differentiable curve denoted by C and f (r) be a continuous vector point function. Then $\frac{dr}{ds}$ is a unit vector function along the tangent at any point P on the curve. The component of the vector function F along this tangent is $F.\frac{dr}{ds}$ which is a function of s for points on the curve.

Then $$\int_C F.\frac{dr}{ds}ds = \int_C F.dr,$$

is called *line integral* or tangent line integral of F(r) along C.

Let $$F = i\,F_1 + j\,F_2 + k\,F_3$$

and $$r = i\,x + j\,y + k\,z$$

$\therefore$ $$dr = i\,dx + j\,dy + k\,dz$$

$\therefore$ $$\int F.dr = \int_C (F_1 dx + F_2 j + F_3 k).(dx\,i + dy\,j + dz\,k)$$

$$= \int (F_1 dx + F_2 dy + F_3 dz)$$

$$= \int \left(F_1 \frac{dx}{dt} + F_2 \frac{dy}{dt} + F_3 \frac{dz}{dt} \right) dt$$

$\therefore$ $$\int_C F.dr = \int_{t1}^{t2} \left(F_1 \frac{dx}{dt} + F_2 \frac{dy}{dt} + F_3 \frac{dz}{dt} \right) dt$$

Where t_1 and t_2 are the values of the parameter t for extremities A and B of the arc of the curve C.

Again, if $$r = x\,i + y\,j + z\,k$$

$\therefore$ $$\frac{dr}{ds} = \frac{dx}{ds}i + \frac{dy}{ds}j + \frac{dz}{ds}k$$

$\therefore$ $$\int_C F.dr = \int_C F.\frac{dr}{ds}ds$$

$$= \int_{s1}^{s2} \left(F_1 \frac{dx}{ds} + F_2 \frac{dy}{ds} + F_3 \frac{dz}{ds} \right) ds$$

Where s_1 and s_2 are the values of s for the extremities of A and B of the arc C.

Physical Interpretation of $\int F.dr$: If F represents a force acting on a particle moving along the curve C then the line integral $\int F.dr$ represents the *work done by the force.* If F represents the velocity of fluid, it is called the *circulation of* F *about* C.

Other Types of Line Integrals

(i) $\int_C F \times dr = \int_C F \times \frac{dr}{ds} ds$

$$= \int_{s1}^{s2} F \times t\, ds$$

where t is a unit tangent vector.

$$\text{Now, } F \times dr = \begin{vmatrix} i & j & k \\ F_1 & F_2 & F_3 \\ dx & dy & dz \end{vmatrix}$$

$$= i(F_2dz - F_3dy) + j(F_3dx - F_1dz) + k(F_1dy - F_2dx)$$

$$\therefore \quad \int_C F \times dr = i\int_C (F_2dz - F_3dy) + j\int_C (F_3dz - F_1dz) + k\int_C (F_1dy - F_2dx)$$

(iii) $\int_C \phi dr = i\int_C \phi dx + j\int_C \phi dy + k\int_C \phi dz,$

where ϕ is a scalar point function.

ILLUSTRATIVE EXAMPLES

1. ***If F = 3x y i − y² j, evaluate*** $\int_C F.dr$***, where C is the curve in the x y-plane y = 2x² from (0, 0) to (1, 2).***

Solution: In the $x\ y$-plane $z = 0$, hence

$$dr = dx\ i + dy\ j$$

$$\therefore \quad \int_C F.dr = \int_C (3xyi - y^2 j).(dxi + dyj)$$

$$= \int_C 3xy\,dx - \int_C y^2 dy$$

Put $y = 2x^2$, $\therefore$ $dy = 4x\,dx$ and x varies from 0 to 1.

$$\therefore \quad \int_C F \times dr = \int_0^1 3x(2x^2)dx - \int_0^1 (2x^2)^2 .4x\,dx$$

$$= \int_0^1 6x^3\,dx - \int_0^1 16x^5\,dx$$

$$= \left[\frac{6}{4}x^4 - \frac{16}{6}x^6\right]_0^1$$

$$= -\frac{7}{6}.$$

2. *Evaluate $\int_C F.dr$, where F = xy i + yz j + zx k and where C is r = i t + j t^2 + k t^3, t varying from –1 to +1.*

Solution: The equation of the curve in parametric form is

$$x = t,\ y = t^2,\ z = t^3$$

$$\therefore \quad F = xy\,i + yz\,j + zx\,k$$

$$= t^2\,i + t^5\,j + t^4\,k$$

Also
$$\frac{dr}{dt} = \frac{dx}{dt}i + \frac{dy}{dt}j + \frac{dz}{dt}k$$

$$= t^3\,i + t^5\,j + t^4\,k$$

$$\therefore \quad F.\frac{dr}{dt} = t^3 + 2t^6 + 3t^6$$

$$= t^3 + 5t^6$$

$$\therefore \quad \int_C F.dr = \int_C F.\frac{dr}{dt}dt$$

$$= \int_{-1}^1 (t^3 + 5t^6)dt$$

$$= \left[\frac{t^4}{4} + \frac{5t^7}{7}\right]_{-1}^1$$

$$= \frac{10}{7}$$

3. Evaluate $\int_C F.dr$***, where*** $F = c[-3\ a\ \sin^2\theta\ \cos\theta\, i + a\ (2\ \sin\theta - 3\sin^3\theta)j + b\ \sin 2\theta\, k]$ ***and is given by*** $r = a\ \cos\theta\, i + a\ \sin\theta\, j + b\theta\, k$ ***for*** $\theta = \pi/4$ ***to*** $\theta = \pi/2$.

Solution: We have

$$\frac{dr}{d\theta} = (-a\sin\theta i + a\cos\theta j + bk)$$

$$\therefore \quad \int_C F.dr = \int_{\pi/4}^{\pi/2} F.\frac{dr}{d\theta}d\theta$$

$$= \int_{\pi/4}^{\pi/2} c\Big[3a^2\sin^3\theta\cos\theta + a^2$$

$$(2\sin\theta\cos\theta - \sin^3\theta\cos\theta) + b^2\sin 2\theta\Big]d\theta$$

$$= \int_{\pi/4}^{\pi/2} c(a^2 + b^2)\sin 2\theta d\theta$$

$$= c(a^2 + b^2)\left[-\frac{\cos 2\theta}{2}\right]_{\pi/4}^{\pi/2}$$

$$= \frac{c}{2}(a^2 + b^2)(-1-0) = \frac{1}{2}c(a^2 + b^2)$$

4. Evaluate $\int_C F.dr$***, where*** $F = (x^2 + y^2)i - 2xyj$ ***and the curve C is the rectangle in the xy plane bounded by*** $y = 0$, $x = a$, $y = b$, $x = 0$.

Solution: In the *x-y*-plane

$$z = 0$$

$$\therefore \quad r = x\,\mathrm{i} + y\,\mathrm{j}$$

or

$$dr = dx\,\mathrm{i} + dy\,\mathrm{j}.$$

$$\therefore \quad \int_C F.dr = \int_C\left[(x^2 + y^2)dx - 2xydy\right] \quad \text{...(i)}$$

Now,

$$\int_C F.dr = \int_O Fdr + \int_{AB} F.dr + \int_{BC} F.dr + \int_{CO} F.dr \quad \text{...(ii)}$$

Along OA, $y = 0$

$\therefore$ $dy = 0$ and x varies from 0 to a.

Along AB, $x = a$

$\therefore$ $dx = 0$ and y varies from 0 to b.

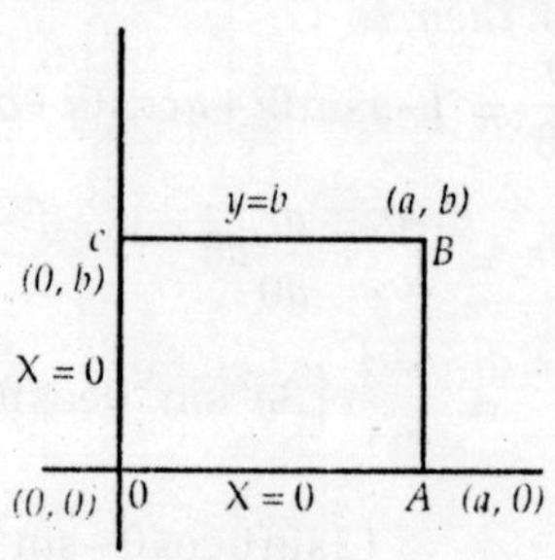

Fig.

Along BC, $y = b$

$\therefore$ $dy = 0$ and x varies from a to 0.

Along CO, $x = 0$

$\therefore$ $dx = 0$ and y varies from b to 0.

Hence, from (i) and (ii), we get

$$\int_C F.dr = \int_0^a x^2 dx - \int_0^b 2ay dy + \int_a^0 (x^2 + b^2) dx + \int_b^0 0.dy$$

$$= \frac{a^3}{3} - 2a.\frac{b^2}{2} + \left[\frac{x^2}{3} + b^2 x\right]_a^0 + 0$$

$$= \frac{a^3}{3} - ab^2 - \frac{a^3}{3} - b^2 a$$

$$= -2ab^2.$$

5. ***If*** $F = (x^2 + y^3)i + (x^3 - y^2)j$, ***evaluate*** $\int F.dr$ ***along the following paths:***

(a) $y^2 = x$, ***joining (0, 0) to (1, 1)***

(b) $x^2 = y$, ***joining (0, 0) to (1, 1)***

(c) Along the straight line joining (0, 0) to (1, 0) and then to (1, 1).

(d) Along the straight line joining (0, 0) to (2, –2) then to (0, –1) and then to (1, 1).

Solution: *Here we have*

$$r = x\,i + y\,j \text{ so that } d\,r = d\,x\,i + d\,y\,j$$

Hence, $$\int_C F.dr = \int_0^1 (x^2 + y^3)dx + (x^3 - y^2)dy \quad \ldots(i)$$

(a) We have $y^2 = x$, $\therefore$ $dx = 2y\,dy$ *and* y *varies from 0 to 1.*

$$\therefore \quad \int_C F.dr = \int_0^1 (y^4 + y^3)(2y\,dy) + (y^6 - y^2)dy$$

$$= \int_0^1 (y^6 + 2y^5 + 2y^4 - y^2)dy$$

$$= \frac{1}{7} + \frac{1}{3} + \frac{2}{5} - \frac{1}{3} = \frac{19}{35}$$

(b) We have $x^2 = y$ $\therefore$ $2x\,dy = dx$ *and* x *varies from 0 to 1.*

$$\therefore \quad \int_C F.dr = \int_0^1 (x^2 + x^6)dx + (x^3 - x^4)2x\,dx$$

$$= \int_0^1 (x^6 - 2x^5 + 2x^4 + x^2)dx$$

$$= \frac{1}{7} + \frac{1}{3} + \frac{2}{5} - \frac{1}{3} = \frac{19}{35}$$

(c) Along the line joining (0, 0) to (1, 0) $y = 0$,

$\therefore$ $d\,y = 0$ and x varies from 0 to 1.

$$\therefore \quad \int_{C1} F.dr = \int_0^1 (x^2 + y^3)dx$$

$$= \int_0^1 x^2\,dx = \frac{1}{3}, \text{ because } y = 0.$$

Along the line (1, 0) to (1, 1) $x = 1$ $\therefore$ $dx = 0$ and y varies from 0 to 1.

$$\therefore \quad \int_{C2} F.dr = \int_0^1 (x^3 - y^2)\,dy$$

$$= \int_0^1 (1 - y^2)\,dy$$

as $\quad x = 1$

$$= \left[y - \frac{y^3}{3}\right] = \frac{2}{3}$$

$$\therefore \quad \int_C F.dr = \int_{C1} F.dr + \int_{C2} F.dr$$

$$= \frac{1}{3} + \frac{2}{3} = 1$$

(d) The equation of the line joining (0, 0) and (2, –2) is

$$y = -x$$

$$\therefore \quad dy = -d\,x \text{ and } x \text{ varies from 0 to 2.}$$

Now put $y = -x$ and $dy = -dx$ in (i) and integrate with in limits 0 to 2.

$$\therefore \quad \int_{C1} F.dr = \int_0^2 (x^2 - x^3)\,dx + (x^3 - x^2)\,dx$$

$$= 2\int_0^2 (x^2 - x^3)\,dx$$

$$= \frac{8}{3}$$

Along C_2 the ling joining (2, –2) to (0, –1) has the equation

$$y + 1 = \frac{-2+1}{2+0}(x - 0)$$

$$y = \frac{(x+2)}{2}$$

$$\therefore \quad dy = \frac{1}{2}dx \text{ and } x \text{ varies from 2 to 0.}$$

Putting the above data in (i) and integrating with respect to x within the above limits, we get

$$\int_{C_2} f.dr = -\frac{9}{2},$$

Similarly,

$$\int_{C_3} f.dr = -\frac{1}{6},$$

$$\therefore \qquad \int_{C} f.dr = -\frac{8}{3}-\frac{9}{2}-\frac{1}{6}=\frac{22}{3}.$$

6. *Find the work done in moving a particle once around a circle C in the x y-plane, if the circle has centre at the origin and radius 2 and if the force field is given by*

$$F = (2x - y + 2z)\, i + (x + y - z^2)j + (3x - 2y - 5z)k$$

Solution: *In the x y-plane,* we have $z = 0$.

$$\therefore \qquad F = (2x - y)i + (x + y)j + (3x - 2y)k$$

C is given by $x^2 + y^2 = 4$

or $\qquad x = 2 \cos t,\ y = 2 \sin t.$

$$\therefore \qquad r = x\, i + x\, j = 2 \cos t\, i + 2 \sin t\, j$$

$$\therefore \qquad \frac{dr}{dt} = -2\sin t\, i + 2 \cos t\, j$$

Also, F = $(4 \cos t - 2 \sin t)\, i + (2 \cos t + 2 \sin t)j + (6 \cos t - 4 \sin t)\, k$.

In going round the circle once, t will very from 0 to 2π.

The work done is

$$\int F.dr = \int_0^{2\pi} F.\frac{dr}{dt}dt$$

$$= \int_0^{2\pi}\left[-2\sin t(4\cos t - 2\sin t) + 2\cos t(2\cos t + \sin t)\right]dt$$

$$= \int_0^{2\pi} 4\left[\left(\sin^2 t + \cos^2 t\right) - 4\sin t + \cos t\, dt\right]$$

$$= \left[4t - 2\sin^2 t\right]_0^{2\pi} = 8\pi.$$

7. Find the work done when a force $F = (x^2 - y^2 + x)i - (2x + y)j$ moves a particle in the xy-plane from (0, 0) to (1, 1) along the parabola $y^2 = x$.

Solution: Work done $\int_C F.dr$

$$= \left[\left(x^2 - y^2 + x\right)dx - \left(2xy + y\right)dy\right] \quad ...(i)$$

Since $y^2 = x$ we may choose $x = t^2, y = t$ and t varies from 0 to 1. Now put $dx = 2tdt, dy = dt$ and change in terms of t.

$\therefore$ Work done $= \int_0^1 \left[2r^5 - \left(2t^3 + t\right)\right]dt$

$$= -\frac{2}{3}.$$

EXERCISE

1. Evaluate $\int_C F.dr$, where F is $x^2y^2i + yj$ and C is $y^2 = 4x$ in the x y-plane from (0, 0) to (4, 4).

2. Evaluate $\int_C F.dr$, where $F = x^2i - xyj$ from point (0, 0) to (1, 1) along the parabola $y^2 = x$.

3. Evaluate $\int_C F.dr$, where $F = \left(x^2 - y^2\right)i + xyj$ and C is the arc of $y = x^3$ from (0, 0) to (2, 8).

4. Evaluate $\int_C F.dr$, where $F = xyi + \left(x^2 + y^2\right)j$ and curve C

is the arc of $y = x^2 - 4$ to from (2, 0) to (4, 12) in the x y-plane.

5. Evaluate $\int_C F.dz,$ where $F = (x^2 + y^2)i + xyj$ and the curve C is the arc of the curve $y = x^2$ from (0, 0) to (3, 9) in the x y-plane.

6. Evaluate $\int_C F.dr,$ where $F = yi - xj$ and C is the arc of the curve $y = x^2$ from (0, 0) to (1, 1).

7. Evaluate $\int_C F.dr,$ where $F = x^2 i + y^2 j + z^2 k$ and C is the arc of the curve $r = ti + t^2 j + t^2 k$ from $t = 0$ to $t = 1$.

8. Evaluate $\int_C F.dr,$ where $F = zi + xj + yk$ in the arc of the curve $r = (\cos t)i + (\sin t)j + tk$ from $t = 0$ to $t = 2\pi$.

9. Evaluate $\int_C F.dr,$ where $F = xyi + yzj + zxk$ and C is the arc of the curve $r = (a\cos\theta)i + (a\sin\theta)j + (a\theta)k$ from $\theta = 0$ to $\theta = \frac{\pi}{2}$.

10. If $F = yz\, i + zxj - xy\, k$ find $\int_C F.dr,$ where C is given by $x = t$, $y = t^2$, $z = t^3$ from $P(0, 0, 0)$ to $Q(2, 4, 8)$.

11. Evaluate $\int_C F.dr,$ where $F = (2x + y)$ i + $(3y - x)$j = yz k and C is the curve $x = 2t^2$, $y = t$, $z = t^3$, from $t = 0$ to $t = 1$.

12. Find $\int_C F.dr,$ where F = (cos y) i – (x sin y) j and C is the curve $y = \sqrt{(1-x)^2}$ in x plane from (1, 0) to (0, 1).

13. If F = $(2x + y)$ i + (3y – x) j, evaluate $\int_C F.dr$ where C is the curve in the x y-plane consisting of straight line from 0(0, 0) to $A(2, 0)$ and then to $B(3, 2)$.

14. If F = $(2y + z)$i + xz j z + $(yz - x)$k, evaluate $\int_C F.dr$ along the following paths C_r,

 (a) $x = 2t^2$, $y = t$, $z = t^3$ from $t = 0$ *to* $t = 1$.

 (b) The straight lines form (0, 0, 0) to (0, 0, 1) then to (0, 1, 1) and then to (2, 1, 1).

 (c) The straight line joining (0, 0, 0) to (2, 1, 1).

15. Find the total work done in moving a particle in a force field given by F = $3xy$ i – 5z j + $10x$ k along the curve $x = t^2 + 1$, $y = 2t^2$, $z = t^3$ from $t = 1$ to $t = 2$.

16. Find the work done in moving a particle once around the curve C is the x y-plane where C is the circle of radius 2 and centre (0, 0) and if the force field is given by F = $3xy$ i – y j + $2zx$ k.

17. Find the work done in moving a particle in the force field F = $3x^2$ i $(2x\ z - y)$ j + zk along

 (a) The line moving (0, 0, 0) to (2, 1, 3).

 (b) The space curve $x = 2t^2$, $y = t$, $z = 4t^2 - t$ from $t = 0$ to $t = 1$.

 (c) The curve defined by $x^2 = 4y$, $3x^2 = 8z$ from $x = 0$ to $x = 2$.

ANSWERS

1. 264.	2. $\frac{1}{12}$.	3. $\frac{824}{21}$.
4. 732.	5. $\frac{774}{5}$.	6. $-\frac{1}{3}$.
7. 1.	8. 3π	9. $a^3\left(\frac{5\pi}{8}-\frac{4}{3}\right)$.

10. $2^7.\frac{5}{42}$. 11. $\frac{277}{42}$. 12. – 1

13. 7

14. (a) $\frac{288}{35}$; (b) 10; (c) 8

15. 303; 16. 0;

17. (a) 16. (b) 14.7; (c) 16.

Normal Surface Integral: Let F(r) be a continuous vector point function and r = f(u, v) possesses continuous first order partial derivatives.

Consider any portion of the surface which may be closed or not.

Divide this surface into a number of subsurfaces SS_1, SS_2, SS_3 and so on. Let δS_p be one of the subsurfaces.

Take any point P_p in this subsurface and let n_p denote the positive unit normal vector to this subsurface at the point P_p.

δS_p is the magnitude of the subsurface and the corresponding vector area be denoted be δa_p.

$$\therefore \quad \delta a_p = n_p \delta S_p$$

Consider the sum $\Sigma F_p \delta a_p = \Sigma F_p . n_p \delta S_p$ ...(i)

The summation extending to various subsurfaces into which S has been divided. Also $F_p . n_p$ denotes the normal component of F_p at P.

The limit of the above sum when the number of subsurfaces tends to infinity and the area of each subsurfaces tends to zero is defined as the *normal surface integral* of F(r) over S and is denoted as

$$\int_S F.d\text{a} = \int_S F.d\text{n}dS.$$

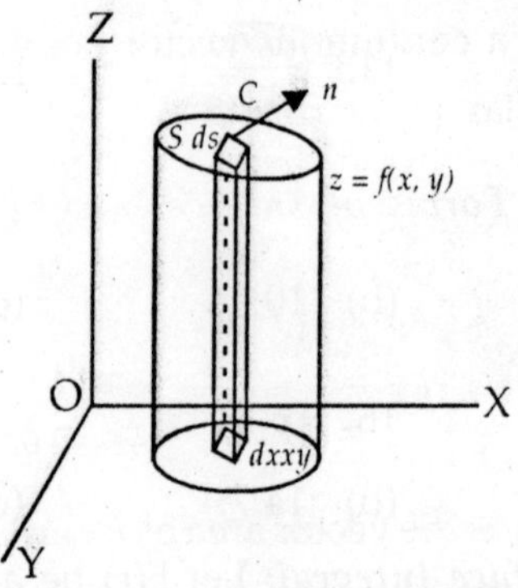

Fig.

The sign of the above integral will change if we choose the normal on the other side.

Cartesian Forms: If F_1, F_2, F_3 be the components of F along the coordinate axes, the

$$\int_S F.nds = \iint_S (F_1 dydz + F_2 dzdx + F_3 dxdy)$$

The above formula can also be put into the form

$$\iint \left[F_1 \frac{\partial(y,z)}{\partial(u,v)} + F_2 \frac{\partial(z,x)}{\partial(u,v)} + F_3 \frac{\partial(x,y)}{\partial(u,v)} \right] dudv$$

Where the integration is to be performed over the region in the u - v plane. Corresponding to the surface S given by

$$r = f(u, v)$$

and $\frac{\partial(y,z)}{\partial(u,v)}$ is the Jacobian $= \begin{vmatrix} \frac{\partial y}{\partial u} & \frac{\partial y}{\partial v} \\ \frac{\partial z}{\partial u} & \frac{\partial z}{\partial v} \end{vmatrix}$, etc.

Other forms of surface integral are

$$\int_S F \times da \text{ and } \int_S \phi da.$$

Where F is a continuous vector point function and ϕ is a continuous scalar point function.

Various Other Forms of Surface Integral

$$\int_S F.da = \int_S F.nds$$

$$= \Sigma F_p \,\delta a_p = \Sigma F_p .n \,\delta S_p$$

Now $n_p \, \delta S_p$ is the vector area of δS_p and hence its projection of x y-plane whose unit normal is k is

$$n_p \, \delta S_p.k = (n_p .k)\delta S_p$$

But projection δS_p on x y-plane is $\delta \times \delta y$

$$\therefore \quad (n_p .k)\delta S_p = \delta \times \delta y$$

$$\therefore \quad \delta S_p = \frac{\partial x \partial y}{n_p.k}$$

$\therefore$ Surface Integral $\int_S F.nds$

$$= \sum F_p.n_p \delta S_p$$

$$= \sum F_p.n_p \frac{\delta x \delta y}{n_p.k}$$

$$= \iint_{S_3} F.n \frac{dx.dy}{n.k}$$

Where S_3 is the projection of S on x y-plane.

Similarly $\int_S F.n dS = \int\int_{S_1} F.n \frac{dydz}{n.i}$

Where S_1 is the projection of S on y z-plane or whose normal is i.

or $$\int_{S_1} F.n\,ds = \iint_{S_2} F.n \frac{dz\,dx}{n.j}$$

Where S_2 is the projection of S on z x-plane whose normal is j.

Volume Integral: Let F (r) be a continuous vector point function and a volume V be enclosed by a surface given by r = f (u, v). Divide the given volume into various $\delta v_1, \delta v_2 ...$ elements. Let δv_p be one such element and P_p be any point on it.

Consider the sum $\Sigma F_p\, \delta v_p$ ...(i)

Where the summation is to be extended to all the elements into which V has been divided. The limit of the above sum when the number of volume elements tends to infinity and each element tends to zero is defined as the *volume integral* and is written as

$$\int_V F dv.$$

In cartesian form

$$\int_V F dv = i \int\int\int_V F_1\, dx\, dy\, dz$$

$$+ \mathrm{j} \int\int\int_V F_2\, dx\, dy\, dz + \mathrm{k} \int\int\int_V F_3\, dx\, dy\, dz$$

ILLUSTRATIVE EXAMPLES

1: Evaluate $\int_S \frac{\mathrm{r}}{r^3} da$, ***where S denotes the spheres of radius a with centre at the origin.***

Solution: Let the equation to the sphere be

$$x^2 + y^2 + z^2 = a^2$$

A normal to the above surface is given by

$$\text{grad}\left(x^2+y^2+z^2\right) = \mathrm{i}(x^2+y^2+z^2)+\mathrm{j}(x^2+y^2+z^2)+\mathrm{k}(x^2+y^2+z^2)$$

$$= 2x\mathrm{i}+2y\mathrm{j}+2z\mathrm{k}$$

$$\therefore \quad \text{Unit normal} = \frac{2x\mathrm{i}+2y\mathrm{j}+2z\mathrm{k}}{\sqrt{\left(4x^2+4y^2+4z^2\right)}}$$

$$= \frac{x\mathrm{i}+y\mathrm{j}+3\mathrm{k}}{a} = \mathrm{n}$$

$$\text{Again} \qquad \mathrm{F} = \frac{\mathrm{r}}{\mathrm{r}^3} = \frac{x\mathrm{i}+y\mathrm{j}+z\mathrm{k}}{\left(x^2+y^2+z^2\right)^{3/2}}$$

$$\frac{x\mathrm{i}+y\mathrm{j}+z\mathrm{k}}{a^3}$$

$$\therefore \quad \int_S \mathrm{F.da} = \int_S \mathrm{F.dn\,dS}$$

$$= \int_S \frac{x\mathrm{i}+y\mathrm{j}+z\mathrm{k}}{a^3}.\frac{x\mathrm{i}+y\mathrm{j}+z\mathrm{k}}{a}dS$$

$$= \int \frac{x^2+y^2+z^2}{a^4}dS = \int_S \frac{a^2}{a^4}dS$$

$$= \frac{1}{a^2}\int_S dS = \frac{1}{a^2}4\pi a^2 = 4\pi$$

2. ***Evaluate*** $\int_S F.ndS$ ***where*** $F = 18z\,\mathbf{i} - 12\mathbf{j} + 3y\,\mathbf{k}$ ***and S is that part of the plane*** $2x + 3y + 6z = 12$ ***which is located in the first octant.***

Solution: We know that grad ϕ is along the normal to the surface $\phi(x,y,z)=0$.

$$\therefore \quad \text{grad}\,\phi = \text{grad}\ (2x + 3y + 6z)$$

$$= 2\mathrm{i} + 3\mathrm{j} + 6\mathrm{k}$$

Hence, n = *a* unit normal to the surface $2x + 3y + 6z = 12$ is

$$= \frac{2i+3j+6k}{\sqrt{(4+9+36)}}$$

$$= \frac{2i+3j+6k}{7}$$

$\therefore$ $$F.a = (18z\ i - 12j + 3y\ k).(2i + 3j + 6k)/7$$

$$= \frac{36z-36+18y}{7}$$

$$= \frac{6(12-2x)-36}{7} = \frac{36-12x}{7}$$

as $2x + 3y + 6z = 12$

$\therefore$ $6z + 3y = 12 - 2x$

$\therefore$ in $x\ y$-plane, we have

$$\int_S F.n\,dS = \iint_{S_3} F.n\frac{dx\,dy}{n.k}$$

Where S_3 is the projection of S on $x\ y$-plane. Now,

$$n.\ k = \frac{2i+3j+6k}{7}$$

$\therefore$ $$\int_S F.ndS = \iint_{S_3} F.n\frac{dx\,dy}{n.k}$$

$$= \iint_{S_3} \frac{36-12x}{7}.\frac{7}{6}dx\,dy$$

$$= \iint(6-2x)dx\,dy.$$

In order to evaluate the above integral the limits of y will vary from $y = 0$ to $y = \frac{12-2x}{3}$ (z = 0 in $x\ y$-plane) and of x will from 0 to 6.

$\therefore$ $$\int_S F.ndS = \int_0^6 \int_0^{12-2x/3} (6-2x)dx\,dy$$

$$= \int_0^6 (6-2x)[y]_0^{12-2x/3} dx$$

$$= \int_0^6 (6-2x)\left(\frac{12-2x}{3}\right)dx$$

$$= \frac{1}{3}\int_0^6 \left(72-36x+4x^2\right)dx$$

$$= \left[24x-6x^2+\frac{4x^3}{9}\right]_0^6$$

$$= 24.$$

3. *If f = y i + (x – 2xz) j – xy k, evaluate* $\int_S (\nabla \times f).\text{n}\,dS$, ***where S is the surface of the sphere $x^2 + y^2 + z^2 = a^2$ above the x y-plane.***

Solution: Let

$$F = \nabla \times \text{f} = \text{curl f} = \begin{vmatrix} \text{i} & \text{j} & \text{k} \\ \dfrac{\partial}{\partial x} & \dfrac{\partial}{\partial y} & \dfrac{\partial}{\partial z} \\ y & x-2x & -xy \end{vmatrix} = x\text{i} + y\text{j} - 2z\text{k}$$

Also, we know that the normal to the surface $x^2 + y^2 + z^2 = a^2$ will be grad $\left(x^2 + y^2 + z^2\right) = 2x\text{i} + 2y\text{j} + 2z\text{k}$

$$\therefore \text{n} = \text{unit normal} = \frac{2x\text{i}+2y\text{j}+2z\text{k}}{\sqrt{\left(4x^2+4y^2+4z^2\right)}}$$

$$= \frac{x\text{i}+y\text{j}+z\text{k}}{a}$$

$$\therefore \quad \text{F. n} = (x\ \text{i} + y\ \text{j} - 2z\text{k}).\left(\frac{x\text{i}+y\text{j}+z\text{k}}{a}\right)$$

$$= \frac{x^2+y^2-2z^2}{a}$$

Also, we know that

$$\int_S F.n dS = \iint_{S_3} F.n \frac{dx\,dy}{n.k}$$

Where S_3 is the projection of S on $x\,y$-plane.

$$n.k = \frac{xi + yj + zk}{a} k = \frac{z}{a}$$

$$= \frac{\sqrt{(a^2 - x^2 - y^2)}}{a}$$

Also, $$F.n = \frac{x^2 + y^2 - 2z^2}{a}$$

$$= \frac{x^2 + y^2 - 2(a^2 - x^2 - y^2)}{a}$$

$$= \frac{3(x^2 + y^2) - 2a^2}{a}$$

$$\therefore \text{ Surface Integral} = \iint_{S_3} \frac{3(x^2 + y^2) - 2a^2}{a} \cdot \frac{dx\,dy}{\sqrt{(a^2 - x^2 - y^2)}} a \quad \ldots\text{(i)}$$

Now, S_3 is the projection of $x^2 + y^2 + z^2 = a^2$ in the $x\,y$-plane and is given by $x^2 + y^2 = a^2$.

In order to integrate (i), put $x = r\cos\theta$, $y = r\sin\theta$

$\therefore$ $d\,x\,d\,y$ - $r\,d\,\theta\,d\,r$ and limits of θ are from 0 to 2π and that of r are form 0 to a.

$$\therefore \quad \int_S F.n dS = \int_0^{2\pi} \int_0^a \frac{3r^2 - 2a^2}{\sqrt{(a^2 - r^2)}} r\,d\theta\,dr$$

$$= 2\pi \int_0^a \frac{3r^2 - 2a^2}{\sqrt{(a^2 - r^2)}} r\,dr$$

Put $a^2 - r^2 = t^2$,

$$\therefore \qquad -2r\,dr = 2t\,dt$$

$$\therefore \qquad \int_S F.ndS = 2\pi \int_0^a \frac{3(a^2 - t^2) - 2a^2}{t}(-t)\,dt$$

$$= 2\pi \int_0^a (a^2 - 3t^2)\,dt$$

$$= 2\pi \left[a^2 t - t^3 \right]_0^a$$

$$= 2\pi (a^3 - a^3) = 0.$$

4. *Evaluate $\int_S F.ndS$, where F = 4x zi – y² j +y z k, where S is the surface of the cube bounded by x = 0, x = 1, y = 0, y = 1, z = 0, z = 1.*

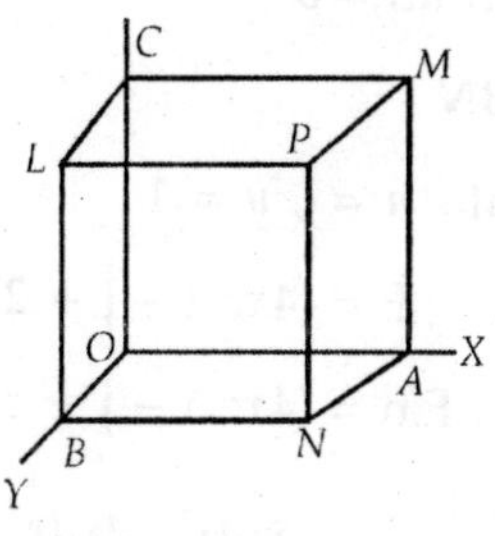

Fig.

Solution: Here we shall calculate $\int F.ndS$ for all the six faces and then add the result.

(i) Face ANPM.

A unit normal n = i and x = 1.

$$\therefore \qquad F = 4z\,i - y^2\,j + yz\,k, \text{ as } x = 1$$

$$\therefore \qquad F\,.\,n = 4z$$

Also $$dS = \frac{dy\,dz}{n.i}$$

$$= \frac{dy\,dz}{\text{i.i}} = dy\,dz$$

$\therefore$ $$\int F.n dS = \int_0^1 \int_0^1 4z\,dy\,dz$$

$$= \int_0^1 \left[2z^2\right]_0^1 dy = \left[2y\right]_0^1 = 2$$

(ii) Face OBLC.

A unit normal n . = – i, $x = 0$

$$F = (-y^2\,j + yz\,k);$$

$\therefore$ $$x = 0;$$

$\therefore$ $$F.n = (-y^2 i + yz\,k).\,(-i) = 0$$

$\therefore$ $$F.n\,dS = 0$$

(iii) Face PLBN

A unit normal $n = j$, $y = 1$

$$F = (4xz\,i - j + 2k); \quad \because y = 1$$

$$F.n = (4xz\,i - j + zk).\,j = -1$$

Also $$dS = \frac{dz\,dx}{\text{n.j}} = \frac{dz\,dx}{\text{j.j}} = dz\,dx$$

$\therefore$ $$\int F.n dS = \int_0^1 \int_0^1 -1 dz\,dx = -1$$

(iv) Face AOCM.

A unit normal $n = -$ j and $y = 0$,

$$F = (4xz\,i + zk); \quad \because F.\,n = 0,$$

$\therefore$ $$F.\,n\,dS = 0.$$

(v) Face MPLC.

A unit normal is k and $z = 1$.

$$\therefore \quad F = (4x\,i - y^2\,j + y\,k), \quad \because z = 1.$$

$$F.n = y.$$

Also $$dS = \frac{dx\,dy}{n.k}dx\,dy$$

$$\therefore \quad \int_S F.ndS = \iint F.n\frac{dx\,dy}{n.k}$$

$$= \int_0^1\int_0^1 y\,dx\,dy = \frac{1}{2}$$

(vi) Face OANB

A unit normal n is –k, z = 0

$$F = -y^2\,j; \quad \because z = 0; \quad \therefore \quad F.n = 0$$

$$\therefore \quad \int_S F.ndS = 0$$

Hence, $$\int_{cube} F.ndS = 2+0-1+0+\frac{1}{2}+0=\frac{3}{2}.$$

5. *If* $F=(2x^2-3z)i-2xyj-4xk$, ***then evaluate*** $\int_V \nabla.FdV$ ***and*** $\iiint_V \nabla\times FdV$, ***where V is the closed region bounded by the plane x = 0, y = 0, z = 0 and 2x + 2y + z = 4.***

Solution: *We have*

$$\nabla.f = \frac{\partial}{\partial x}(2x^2-3z)+\frac{\partial}{\partial y}(-2xy)+\frac{\partial}{\partial z}(-4x)$$

$$= 4x - 2x - 0 = 2x$$

and $$dV = dx\,dy\,dz.$$

Limits of z are from 0 to 4 – (2x + 2y), limits of y are from 0 to 2 – x and limits of x are from 0 to 2.

$$\therefore \quad \int_V = \nabla.fdV$$

$$= \int_0^{4-(2x+2y)} \int_0^{(2-x)} \int_0^{2} 2x\,dx\,dy\,dz$$

$$= \int_0^{(2-x)} \int_0^{2} 2x(4-2x-2y)\,dx\,dy$$

$$= \int_0^{2} \left[8xy - 4x^2y - 2xy^2\right]_0^{2-x} dx$$

$$= \int_0^{2} \left[8x(2-x) - 4x^2(2-x) - 2x(2-x)^2\right] dx$$

$$= \int_0^{2} \left(2x^3 - 8x^2 + 8x\right) dx = \left[2.\frac{2^4}{4} - 8.\frac{2^3}{3} + 8.\frac{2^2}{2}\right]$$

$$= \frac{8}{3}.$$

Again, $\nabla \times F = \begin{vmatrix} i & j & k \\ \dfrac{\partial}{\partial x} & \dfrac{\partial}{\partial y} & \dfrac{\partial}{\partial z} \\ 2x^2 - 3z & -2xy & -4x \end{vmatrix}$

$= j - 2ky$

$$\therefore \int_V \nabla \times \mathrm{f}\,dV = \iiint (j - 2ky)\,dx\,dy\,dz$$

$$= \int_0^{(4-2x-2y)} \int_0^{2-x} \int_0^{2} (j - 2ky)\,dx\,dy\,dz$$

$$= \int_0^{2} \left[j(4y - 2xy - y^2) - 2k\left(2y^2 - xy^2 - \frac{2y^3}{3}\right)\right]_0^{2-x} dx$$

$$= \int_0^{2} \left[j(2-x)(4-2x-2+x) - 2k(2-x)^2\left\{2 - x - \frac{2}{3}(2-x)\right\}\right] dx$$

$$= \int_0^{2} \left[j(2-x)^2 - \frac{2k}{3}(2-x)^3 \right] dx$$

$$= \left| j\frac{(x-2)^3}{3} + \frac{2k}{3}\frac{(x-2)^4}{4} \right|_0^2$$

$$= j.\frac{8}{3} + \frac{2k}{3}\left(-\frac{16}{4}\right) = \frac{8}{3}(j - k)$$

6. *Evaluate* $\int_V (2x+y)dv$, *where V is the closed region bounded by the cylinder z = 4 – x² and the planes x = 0, y = 0, y = 2 and z = 0.*

Solution: The given cylinder $z = 4 - x^2$ meets the x-axis $y = 0, z = 0)$ at $x^2 = 4$ or $x = 2$, *i.e.*, at the point (2, 0, 0). It meets z-axis $(x, = 0, y = 0)$ at $z = 4$, *i.e.*, at (0, 0, 4). Therefore the limits of integration are:

$$z = 4 \text{ to } z = 4 - x^2,$$

$$y = 0 \text{ to } y = 2$$

and $$x = 0 \text{ to } x = 2.$$

Also $$dV = dx\, dy\, dz$$

$$\therefore \int_V (2x-y)dv = \int_{x=0}^{2}\int_{y=0}^{2}\int_{z=0}^{(4-x)^2} (2x+y)\,dx\,dy\,dz$$

$$= \int_{x=0}^{2}\int_{y=0}^{2}(2x+y)[z]_0^{4-x^2}\,dx\,dy$$

$$= \int_{x=0}^{2}\int_{y=0}^{2}(2x+y)(4-x^2)\,dx\,dy$$

$$= \int_0^2\left[2x(4-x^2)y+(4-x^2)\frac{y^2}{2}\right]_{y=0}^{2} dx$$

$$= \int_0^2\left[4x(4-x^2)+2(4-x^2)\right]dx$$

$$= \int_0^2 2(4-x^2)(1+2x)dx$$

$$= 2\int_0^2 (4+8x-x^2-2x^3)dx$$

$$= 2\left[4x+4\times 2-\frac{x}{3}-\frac{x^4}{2}\right]_0^2$$

$$= 2\left(8+16-\frac{8}{3}-8\right)=\frac{80}{3}.$$

EXERCISE

1. Evaluate $\int_S \text{f}.n dS$ where f = y i + 2x j + zk and S is the surface of the plane 2x + y = 6 in the first octant cut off by the plane z = 4.

2. Evaluate $\int_S \text{f}.n dS$ over the surface of the cylinder $x^2 + y^2 = 9$ included in the first octant between z = 0 and z = 4 where f = z i + x j – yz k.

3. Evaluate $\int_S \text{f}.n dS$ where f = 4x i – 2y^2 j + z^2 k taken over the region bounded x^2 + y^2 = 4, z = 0 and z = 3.

4. Evaluate $\int_S \text{f}.n dS$ where F = 2xy i – zy j + x^2 k over the surface of cube bounded by the coordinate planes and the planes x = z, y = z, z = a.

5. Evaluate $\int_S (x^2\text{i} + y^2\text{j} + z^3\text{k}).da$, where S is the surface of the sphere $x^2 + y^2 + z^2 = 1$.

6. Let $\phi = 45x^2y$ and let V denote the closed region bounded by planes $4x + 2y + z = 8$. $y = 0, z = 0$. Evaluate, $\int_V \phi dV$.

7. Evaluate $\int_V \text{f} dV$ for f = x i + y j + z k where V is the region bounded by the surface x = 0, y = 6, z = 4 and z = x^2.

ANSWERS

1. 108
2. 42
3. 84 π
4. $\frac{1}{2}a^4$
5. $\frac{12}{5}\pi$
6. 128
7. $24\text{i} + 96\text{j} + \frac{384}{5}\text{k}$

Gauss' Divergence Theorem

This theorem states that the surface integral of the normal component of a vector $\vec{A}$ taken over a closed surface S is equal to the volume integral of the divergence of $\vec{A}$ taken over the volume V enclosed by the surface S, *i.e.*,

$$\int_S \int \vec{A}.d\vec{S} = \int\int_V \text{div}\,\vec{A}dV.$$

Thus, the theorem gives us a method of reducing triple integrals to double integrals.

Proof: let us consider a 'closed' surface S of any arbitrary shape drawn in a vector field $\vec{A}$ and enclosing a volume V. Let dV be a small cubical volume element. The flux which diverges from this element is div $\vec{A}dV$, because div $\vec{A}$ represents the amount of flux diverging per unit volume. The total flux coming out from the entire volume V is therefore

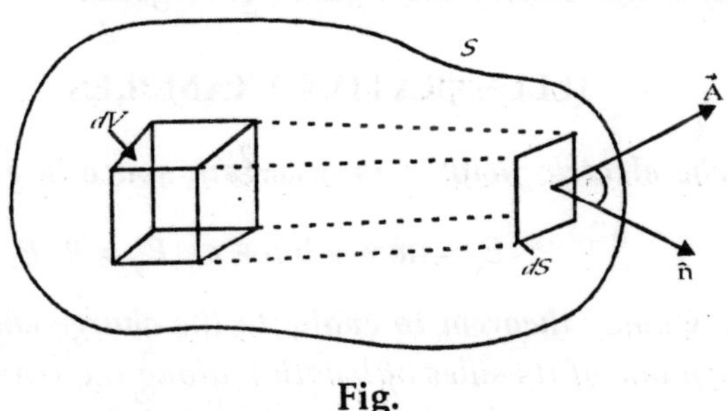

Fig.

$$\int\int_V \int \text{div}\,\vec{A}dV. \quad \text{...(i)}$$

Let us now consider a small element of area dS upon the surface S. Let $\hat{n}$ be a unit positive (outward) normal upon dS. If θ be the angle between $\vec{A}$ and $\hat{n}$, then the component of $\vec{A}$ along $\hat{n}$ is

$$A\cos\theta = \vec{A}.\hat{n}.$$

The flux of $\vec{A}$ through the surface element dS, defined as the product of the normal component of $\vec{A}$ and the area dS, is

$$\left(\vec{A}.\hat{n}\right)dS = \vec{A}.d\vec{S}$$

The total flux through the entire surface S is therefore

$$\iint_S \vec{A}.d\vec{S} \qquad ...(ii)$$

This must be the same as the total flux diverging from the volume V enclosed by the surface S. Thus, (i) and (ii) are equal, *i.e.*,

$$\iint_S \vec{A}.d\vec{S} = \iiint_v \text{div}\,\vec{A}.dV.$$

This may also be written as

$$\iint_S \vec{A}.\hat{n}\,dS = \iiint_v (\vec{\nabla}.\vec{A}).dV$$

This is the Gauss' theorem of divergence.

ILLUSTRATIVE EXAMPLES

1. The electric field $\vec{E}$ in a certain space is given by

$$E_x = (a\,x + by + c),\ E_y = 0,\ E_z = 0.$$

Use Gauss' theorem to evaluate the charge enclosed in a cube with one of its sides of length L along the x-axis, the two faces of the cube perpendicular to the x-axis lying in the planes x = l and x = l + L.

Solution: According to the Gauss's theorem of divergence, the surface integral of the normal component of electric field $\vec{E}$ taken over a closed surface S is equal to the volume integral of div $\vec{E}$ taken over the volume V enclosed by the surface S. That is,

$$\int_S\int \vec{E}.dS = \iint_V \int \text{div}\,\vec{E}\,dV$$

$$= \iiint_V \left(\frac{\partial E_x}{\partial x} + \frac{\partial E_y}{\partial y} + \frac{\partial E_z}{\partial z} \right) dV$$

Here $\frac{\partial E_x}{\partial x} = a$, while E_y and E_z is each zero. Therefore,

$$\iint_S \vec{E}.d\vec{S} = \iiint_V a\,dV$$

$$= aV = aL^3 \qquad \text{...(i)}$$

Since the volume V of the cube L^3. But, by Gauss' law in electrostatics, we have

$$\iint_S \vec{E}.d\vec{S} = \frac{q}{\varepsilon_0},$$

where q is the charge within the surface S which encloses the volume V. Equating (ii) and (i), we have

$$\frac{q}{\varepsilon_0} = aL^3$$

$$\therefore \qquad q = \varepsilon_0 a L^3 .$$

2. The electric field at any point in a given space is directed radically outward along the line joining the point to a fixed point ($r = 0$) and is given by $a\ r + b\ r^2$. Evaluate the charge enclosed in a sphere of radius R with the fixed point ($r = 0$) as centre.

Solution: The magnitude of the radial electric field is given by

$$E = a\ r + b\ r^2$$

Its value at the surface of the sphere of radius R is therefore

$$E = a\ R + b\ R^2$$

The surface integral of the normal component of $\vec{E}$ over the closed surface S of the sphere is

$$\int_S \int \vec{E}.d\vec{S} = \iint_S E\,dS$$

$$[\because \vec{E} \text{ and } d\vec{S} \text{ are in same direction}]$$

$$= \int_S\int\left(aR+bR^2\right)dS$$

$$= \left(aR+bR^2\right)\int\int_S dS$$

$$= \left(aR+bR^2\right)\left(4\pi R^2\right)$$

$$= 4\pi R^3\left(a+bR\right) \quad \ldots\text{(i)}$$

But by Gauss' law in electrostatics, we have

$$\int_S\int \vec{E}.d\vec{S} = \frac{q}{\varepsilon_0}, \quad \ldots\text{(ii)}$$

where q is the charge enclosed by the surface S. Equating (ii) and (i), we get

$$\frac{q}{\varepsilon_0} = 4\pi R^3\left(a+bR\right), \quad \therefore\ q = 4\pi\varepsilon_0 R^3\left(a+bR\right).$$

3. *Evaluate $\int_S F.ndS$, where F = 4xz i – y^2 j – yz k and S is the surface of the cube bounded by x = 0, x = 1, y = 0, y = 1, z = 0 and z = 1.*

Solution: By Gauss's divergence theorem, we have

$$\int_S F.ndS = \int_V div\, F\, dV$$

We have

$$F = 4xz\ i - y^2\ j + yz\ k$$

$$\therefore \quad \text{div } F = \frac{\partial}{\partial x}(4xz)+\frac{\partial}{\partial y}(-y^2)+\frac{\partial}{\partial z}(yz)$$

$$= 4x - 2y + y = 4z - y$$

Hence, by divergence theorem, we get

$$\int_S F.ndS = \int_{x=0}^{1}\int_{y=0}^{1}\int_{z=0}^{1}(4z-y)dx\,dy\,dz$$

$$= \int_{x=0}^{1}\int_{y=0}^{1}\left[2z^2 - yz\right]_{z=0}^{1} dx\,dy$$

$$= \int_{x=0}^{1}\int_{y=0}^{1}(2-y)\,dx\,dy$$

$$= \int_0^1\left[2y - \frac{y^2}{2}\right]_0^1 dx$$

$$= \frac{3}{2}\int_0^1 dx = \frac{3}{2}.$$

4. ***If F = 2xy i– yz j + x² k, evaluate*** $\int_S F.ndS$, ***where S denotes the entire surface of the cube bounded by the coordinate planes and the planes x = a, y =a, z = a by the application of Gauss's theorem***

Solution: We have

$$F = 2xy\ i - yz\ j + x^2\ k$$

$$\therefore \quad \text{div } F = \frac{\partial}{\partial x}(2xy) + \frac{\partial}{\partial y}(-yz) + \frac{\partial}{\partial z}(x^2)$$

$$= 2y - z$$

$$\therefore \quad \int_S F.ndS = \int_V \text{div } F\,dV,$$

by Gauss's divergence theorem

$$= \int_0^a\int_0^a\int_0^a (2y - z)\,dx\,dy\,dz$$

$$= \int_0^a\int_0^a\left[2yz - \frac{z^2}{2}\right]_0^a dx\,dy$$

$$= \int_0^a\int_0^a\left(2ay - \frac{a^2}{2}\right)dx\,dy$$

$$= \int_0^a\left[ay^2 - \frac{1}{2}a^2y\right]_0^a dx$$

$$= \int_0^a \left(a^3 - \frac{1}{2}a^3 \right) dx = \frac{1}{2}a^3 [x]_0^a$$

$$= \frac{1}{2}a^4.$$

5. Prove that $\int_S n.(\nabla \times F) dS = 0$, ***where F is a vector point function and S is a closed surface.***

Solution: Let f = $\nabla \times F$ = curl F

$$\therefore \quad \int_S n.(\nabla \times F) ds = \int_S n.f dS$$

$$= \int_V \text{div } f dV,$$

by Gauss's Divergence Theorem

$$= \int_V \text{div}(\text{curl F}) dV = 0$$

because div (curl f) = 0

6. Show that $\int F.n dS = \int A^2 dV$ ***where*** $F = \phi A; A = \nabla\phi$ ***and*** $\nabla^2 \phi = 0$.

Solution: We have

$$\text{div F} = \text{div } (A\ \phi) = \phi \,\text{div}\, A + \nabla\phi . A$$

$$= \phi \nabla.(\nabla\phi) + A.A$$

$$= \phi \nabla^2 \phi + A^2 = A^2,$$

because $\nabla^2 \phi = 0$

Hence, by Gauss's Divergence Theorem

$$\int_S F.n dS = \int_S \text{div} F dV$$

$$= \int A^2 dV.$$

7. Verify divergence theorem for F = x i − yj + (z² − 1)k taken over the region bounded by x² + y² = 4, z = 0, z = 1.

Solution: We have, Gauss's divergence theorem as

$$\int_S F.ndS = \int_S \text{div}.FdV$$

Now, $$\text{div } F = \frac{\partial}{\partial x}(x) + \frac{\partial}{\partial y}(-y) + \frac{\partial}{\partial z}(z^2 - 1)$$

$$= 1 - 1 + 2z = 2z.$$

$$\therefore \quad \int_V \text{div}\, F\, dV = \int_{-2}^{2}\int_{\sqrt{(4-x2)}}^{\sqrt{(4-x2)}}\int_0^1 2z\,dx\,dy\,dz$$

$$= \int_{-2}^{2}\int_{\sqrt{(4-x2)}}^{\sqrt{(4-x2)}} \left[z^2\right]_0^1 dx\,dy$$

$$= 2\int_{-2}^{2}\int_0^{\sqrt{(4-x2)}} dx\,dy$$

$$= 2.2\int_0^2 \sqrt{(4-x^2)}\,dx$$

$$= 4.\left[\frac{x}{2}\sqrt{(4-x^2)} + \frac{4}{2}\sin-1\frac{x}{2}\right]_0^2$$

$$= \left[0 + 2.\frac{\pi}{2}\right] = 4\pi.$$

The surface integral can be evaluated as given in previous part.

8. Show that $\int_S F \times ndS,$, *i.e.,* $\int_S F \times da = -\int_V curl\, FdV.$

Solution: Let f = a × F, where a is a constant vector.

$$\therefore \quad \int_S f.ndS = \int_V \text{div}\, f\, dV$$

or $$\int_S \text{a} \times \text{F. n}dS = \text{div (a} \times \text{F)}\, dV \qquad \text{...(i)}$$

Now,

$$a \times F \,.\, n = a \,.\, F \times n, \quad \text{...(ii)}$$

because, position of dot and cross can be changed in scalar triple product if cyclic order is maintained.

Also, $\text{div } (a \times F) = (\text{curl a}) \,.\, F - (\text{curl f}) \,.a$

$$= -a \,.\, (\text{curl F}), \quad \text{...(iii)}$$

as $\quad$ curl a = 0.

Substituting the above results (ii) and (iii) in (i), we get

$$\int_S a \,.\, F \times n\, dS = -a.\ \text{curl F } dV$$

or $\quad a.\left[\int_S (F \times n)\, dS + \int_V \text{curl F}\, dV\right] = 0 \quad \text{...(iv)}$

Since a is any arbitrary vector, we have from (iv),

$$\int_S F \times n\ dS + \int_V \text{curl F } dV = 0$$

$$\therefore \quad \int_S F \times n\ dS = -\int_V \text{curl F } dV.$$

9. Prove that $\int \phi n dS$,, ***i.e.,*** $\int \phi da = \int_1 \nabla \phi dV$.

Solution: Let us choose $f = a\phi$, where **a** is a constant vector then by Gauss's Divergence theorem.

$$\int_S f.n dS + \int_V (\text{div f}) dV$$

or
$$\int_S a.\phi.n dS = \int_V \text{div}(a\phi) dV$$

$$= \phi \,\text{div}\, a + a.\nabla\phi$$

$$= a.\nabla\phi, \text{ as div } a = 0$$

or $$a.\int_S \phi n dS = \int_V a.\nabla\psi dV$$

or $$\left[a.\int_S \phi n dS = \int_V a.\nabla\phi dV\right] = 0 \qquad \ldots(i)$$

Since a is an arbitrary vector, we have, from (i)

$$\int_S \phi dS - \int_V \nabla\phi dV = 0$$

$$\therefore \int_S \phi n dS - \int_V \nabla\phi dV.$$

EXERCISE

1. Show that $\frac{1}{3}\int_S r.n dS = V.$
2. Evaluate $\int_S F.n dS$ when f = $4xy$ i + yz j – xz k and S is the surface of the cube bounded by the plane $x = 0$, $x = 2$, $y = 0$ and $z = 0$, $z = 2$.
3. If V is the volume enclosed by a closed surface S and F = x i + $2y$ j + $3z$ k, show that $\int_S F.n dS = 6V.$
4. Evaluate $\int_S (xi + yj + zk).n ds$ where S denotes the surface of the cube bounded by the planes $x = 0$. $x = a$, $y = 0$, $y = a$, $z = 0$, $z = a$ by the application of Gauss's Theorem.
5. Verify divergence theorem for f = $(2x - z)$i + x^2y j – xz^2 k, taken over the region bounded by $x = 0$, $x = 1$, $y = 0$, $y = 1$, $z = 0$, $z = 1$.
6. Verify the divergence theorem for f = $4x$ i – $2y^2$ j + z^2 k taken over the region bounded by $x^2 + y^2 = 4$, $z = 0$ *and* $z = 3$.
7. Verify the divergence theorem for the function F = y i + x j + z^2 k over the cylindrical region S bounded by $x^2 + y^2 = a^2$, $z = 0$ *and* $z = h$.

8. Evaluate $\int_S (y^2z^2\mathrm{i}+z^2x^2\mathrm{j}+z^2y^2\mathrm{k}).\mathrm{n}\,dS$, where S is the part of the sphere $x^2+y^2+z^2=1$ above the x y-plane.

9. Show that $(x^2\mathrm{i}+y^2\mathrm{j}+z^2\mathrm{k}).\mathrm{n}\,dS=0$ where S denotes the surface of the ellipsoid $\frac{x^2}{a^2}+\frac{y^2}{b^2}+\frac{z^2}{c^2}=1.$

10. Evaluate $\int_S (x\mathrm{i}+y\mathrm{j}+z^2\mathrm{k}).\mathrm{n}dS,$ where S is the closed surface bounded by the cone $x^2+y^2=z^2$ and the plane $z = 1$.

11. Apply Gauss's Divergence theorem to show that $(x^2\mathrm{i}+z\mathrm{j}+yz\mathrm{k}).\mathrm{n}\,dS=\frac{3}{2}$, where S is the unit cube bounded by $x = 0$, $x = 1$, $y = 0$, $y = 1$, $z = 0$, $z = 1$.

Green's Theorem

If ϕ and ψ are two scalar point functions such that these functions and their first derivatives are continuously differentiable, then we have

$$\int_V\int\int(\phi\nabla^2\psi+\vec{\nabla}\phi.\vec{\nabla}\psi)dV = \int_S\int(\phi\vec{\nabla}\psi).d\vec{S},$$

and
$$\int_V\int\int(\phi\nabla^2\psi-\psi\nabla^2\phi)dV = \int_S\int(\phi\vec{\nabla}\psi-\psi\vec{\nabla}\phi).d\vec{S}.$$

These equations are respectively known as first and second form of Green's theorem.

Proof: Let us start with Gauss's divergence theorem which states

$$\int_V\int\int \mathrm{div}\vec{A}\,dV = \int_S\int\vec{A}\,d\vec{S}. \qquad \ldots\text{(i)}$$

Let us put

$$\vec{A} = \phi\vec{\nabla}\psi$$

or $$A_x\hat{i}+A_y\hat{j}+A_z\hat{k} = \phi\left(\frac{\partial\psi}{\partial x}\hat{i}+\frac{\partial\psi}{\partial y}\hat{j}+\frac{\partial\psi}{\partial z}\hat{k}\right)$$

so that $$A_x = \phi\frac{\partial\psi}{\partial x}, A_y = \phi\frac{\partial\psi}{\partial y}, A_z = \phi\frac{\partial\psi}{\partial z}.$$

Now, div $$\vec{A} = \frac{\partial A_x}{\partial x}+\frac{\partial A_y}{\partial y}+\frac{\partial A_z}{\partial z}$$

$$= \frac{\partial}{\partial x}\left(\phi\frac{\partial\psi}{\partial x}\right)+\frac{\partial}{\partial y}\left(\phi\frac{\partial\psi}{\partial y}\right)+\frac{\partial}{\partial z}\left(\phi\frac{\partial\psi}{\partial z}\right)$$

$$= \left(\phi\frac{\partial^2\psi}{\partial x^2}+\frac{\partial\phi}{\partial x}\frac{\partial\psi}{\partial x}\right)+\left(\phi\frac{\partial^2\psi}{\partial y^2}+\frac{\partial\phi}{\partial y}\frac{\partial\psi}{\partial y}\right)+\left(\phi\frac{\partial^2\psi}{\partial z^2}+\frac{\partial\phi}{\partial z}\frac{\partial\psi}{\partial z}\right)$$

$$= \phi\left(\frac{\partial^2\psi}{\partial x^2}+\frac{\partial^2\psi}{\partial y^2}+\frac{\partial^2\psi}{\partial z^2}\right)+\frac{\partial\phi}{\partial x}\frac{\partial\psi}{\partial x}+\frac{\partial\phi}{\partial y}\frac{\partial\psi}{\partial y}+\frac{\partial\phi}{\partial z}\frac{\partial\psi}{\partial z}$$

$$= \phi\nabla^2\psi+\vec{\nabla}\phi.\vec{\nabla}\psi.$$

Making this substitution in the divergence theorem (i). we get,

$$\iiint_V \left(\phi\nabla^2\psi+\vec{\nabla}\phi.\vec{\nabla}\psi\right)dV = \iint_S \left(\phi\vec{\nabla}\psi\right).d\vec{S} \qquad \text{...(ii)}$$

This equation is referred to as "Green's first theorem".

Interchanging ϕ and ψ in Eq. (ii), we get

$$\iiint_V \left(\psi\nabla^2\phi+\vec{\nabla}\psi.\vec{\nabla}\phi\right)dV = \iint_S \left(\psi\vec{\nabla}\phi\right).d\vec{S}. \qquad \text{...(iii)}$$

Subtracting (iii) from (ii), we get

$$\iiint_V (\phi\nabla^2\psi - \psi\nabla^2\phi)dV = \iint_S (\phi\vec{\nabla}\psi - \psi\vec{\nabla}\phi).d\vec{S} \quad \text{...(iv)}$$

This equation is referred to as "Grean's second or symmetrical theorem". Now, in the R.H.S., we have

$$\vec{\nabla}\psi.d\vec{S} = \frac{\partial\psi}{\partial n}dS.$$

and
$$\vec{\nabla}\phi.d\vec{S} = \frac{\partial\phi}{\partial n}dS,$$

where $\frac{\partial\psi}{\partial n}$ and $\frac{\partial\phi}{\partial n}$ are the directional derivatives of ψ and ϕ, respectively along the outward normal to dS. Therefore, Eq. (iv) can also be written as

$$\iiint_V (\phi\nabla^2\psi - \psi\nabla^2\phi)dV = \iint_S \left(\phi\frac{\partial\psi}{\partial n} - \psi\frac{\partial\phi}{\partial n}\right)dS.$$

This is an alternative form of Green's symmetrical theorem.

Green's Theorem in a Plane

It states that if R is a closed region in the $x\,y$-plane bounded by a closed curve C, and if M and N are continuous functions of x and y having continuous derivatives in R, then

$$\oint_C (Mdx + Ndy) = \iint_R \left(\frac{\partial N}{\partial x} - \frac{\partial M}{\partial y}\right)dx\,dy,$$

where C is traversed in the positive (counter clockwise) direction.

Proof: Let us consider a closed curve C bounding a region R in the x-y plane. The curve has the property that any straight line parallel to the coordinate axes cuts the curve in at most two points. Let us choose on the curve points A, B, E and F such that the tangents at these points are parallel to the coordinate axes. Let $y = f_1(x)$ and $y = f_2(x)$ be the equations of the curves AEB and AFB, respectively. Then we have,

$$\int_R \int \frac{\partial M}{\partial y} dx\, dy = \int_{x=a}^{b} \left[\int_{y=f_1(x)}^{f_2(x)} \frac{\partial M}{\partial y} \partial y \right] dx$$

$$= \int_{x=a}^{b} \left[M(x,y) \right]_{y=f_1(x)}^{f_2(x)} dx$$

$$= \int_a^b \left[M(x,y_2) - M(x,y_1) \right] dx$$

where $y_2 = f_2(x)$

and $y_1 = f_1(x)$

$$= \int_a^b M(x,y_2) dx - \int_a^b M(x,y_1) dx$$

$$= -\int_b^a M(x,y_2) dx - \int_a^b M(x,y_1) dx$$

$$= -\left[\int_a^b M(x,y_1) dx + \int_b^a M(x,y_2) dx \right]$$

$$= -\oint_C M dx.$$

From this, we have

$$= -\oint_C M dx = -\int_R \int \frac{\partial M}{\partial y} dx\, dy \qquad \text{...(i)}$$

Similarly, let $x_1 = f_3(y)$ and $x_2 = f_4(y)$ be the equations of the curves *EAF* and *EBF*, respectively. Then we have

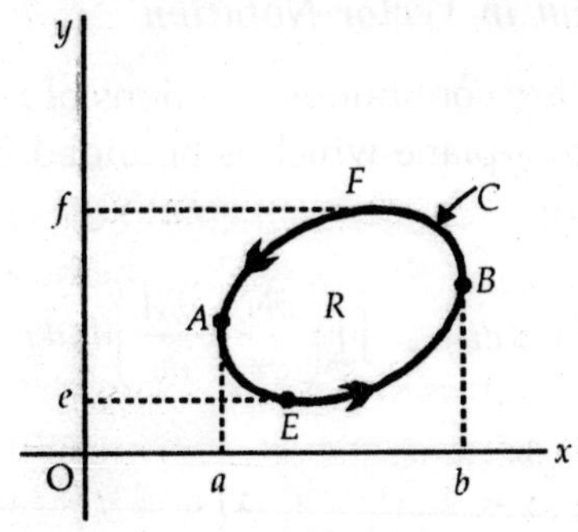

Fig.

$$\iint_R \frac{\partial N}{\partial x} dx\, dy = \int_{y=e}^{f}\left[\int x_{f_3(y)}^{f_4(y)} \frac{\partial N}{\partial x} dx\right] dy$$

$$= \int_{y=e}^{f}\left[N(x,y)\right] x =_{f_3(y)}^{f_4(y)} dy$$

$$= \int_{e}^{f}\left[N(x_2,y) - N(x_1,y)\right] dy;$$

where $x_2 = f_4(y)$

and $x_1 = f_3(y)$

$$= \int_E^F N(x_2,y)dy - \int_E^F N(x_1,y)dy$$

$$= \int_F^E N(x_1,y)dy + \int_E^F N(x_2,y)dy$$

$$= \oint_C N\, dy.$$

From this, we have

$$\oint_C N\, dy = \iint_R \frac{\partial N}{\partial x} dx\, dy \qquad \text{...(ii)}$$

Adding (i) and (ii), we get

$$\oint_C (M\, dx + N\, dy) = \iint_R \left(\frac{\partial N}{\partial x} - \frac{\partial M}{\partial y}\right) dx\, dy. \qquad \text{...(iii)}$$

This is Green's theorem in the plane.

Green's Theorem in Vector Notation

If M and N are continuous functions of x and y in a closed region R in the xy-plane which is bounded by a closed curve C, then we have

$$\oint_C (M\, dx + N\, dy) = \iint_R \left(\frac{\partial N}{\partial x} - \frac{\partial M}{\partial y}\right) dx\, dy \qquad \text{...(i)}$$

Let us put

$$\vec{A} = M\hat{\imath} + N\hat{\jmath}.$$

Also, we have in $x\ y$-plane

$$\vec{r} = x\hat{i} + y\hat{j}$$

so that
$$d\vec{r} = dx\hat{i} + dy\hat{j}$$

Now,
$$Mdx + Ndy = (M\hat{i} + N\hat{j}).(dx\hat{i} + dy\hat{j})$$
$$= \vec{A}.d\vec{r} \qquad \text{...(ii)}$$

Further,
$$\text{curl}\,\vec{A} = \begin{vmatrix} \hat{i} & \hat{j} & \hat{k} \\ \dfrac{\partial}{\partial x} & \dfrac{\partial}{\partial y} & \dfrac{\partial}{\partial z} \\ M & N & O \end{vmatrix}$$
$$= \hat{i}\left(0 - \frac{\partial N}{\partial z}\right) - \hat{j}\left(0 - \frac{\partial M}{\partial z}\right) + \hat{k}\left(\frac{\partial N}{\partial x} - \frac{\partial M}{\partial y}\right)$$
$$= -\frac{\partial N}{\partial z}\hat{i} + \frac{\partial M}{\partial z}\hat{j} + \left(\frac{\partial N}{\partial x} - \frac{\partial M}{\partial y}\right)\hat{k}$$

so that
$$\text{curl}\,\vec{A}.\vec{K} = \left[\frac{\partial N}{\partial z}\hat{i} + \frac{\partial M}{\partial z}\hat{j} + \left(\frac{\partial N}{\partial x} - \frac{\partial M}{\partial y}\right)\hat{k}\right].\hat{k}$$
$$= \frac{\partial N}{\partial x} - \frac{\partial M}{\partial y} \qquad \text{...(iii)}$$

Putting the value of $M\,dx + N\,dy$ from (ii) and the value of $\dfrac{\partial N}{\partial x} - \dfrac{\partial M}{\partial y}$ from (iii) in Eq. (i), we get

$$\oint_C \vec{A}.d\vec{r} = \iint_R \text{curl}\,\vec{A}.\hat{k}\,dR,$$

where $d\,R = dx\,dy$. This is Green's theorem in vector notation.

Different Operations

Importance of Fourier Series

Consider a function $f(t)$ of a variable t. $f(t)$ is called a periodic function of t if it repeats itself regularly over a given interval of t (*i.e.* t may be a time variable or a space variable). Mathematically, a function $f(t)$ is said to be periodic if for all t, there is some finite positive T such that

$$f(t + T) = f(t), \qquad \text{...(1)}$$

T is called the period of $f(t)$. If r is any integer, then it is clear from the above relation that

$$f(t + rT) = f(t), \qquad \text{...(2)}$$

for all t. Hence, rT , *i.e.* $2T, 3T, 4T$ ... are also periods of $f(t)$.

sin t, cos t are periodic functions having a period 2π, tan t has a period of π , and sin rt and cos rt have periods $2\pi/r$.

Equal Works

A function $f(t)$ is an even function if $f(-t) = f(t)$

Examples: t^{2n}, *cos t, t sin t are even function of t.*

Unequal Works

A function $f(t)$ is said to be an odd function of t

if $f(-t) = -f(t)$.

examples t^{2n+1} (for integer n), $\sin t$, $t \cos t$ are odd functions of t.

Fourier Series

In science and engineering we sometimes come across a complicated function $f(t)$ of period say $T = 2\pi$. We also know that $\sin t$, $\sin 2t$, $\cos t$, $\cos 2t$, ... also are periodic functions with a period 2π.

The question arises," under what conditions can we express $f(t)$ of period 2π in terms of combination of $\sin rt$ and $\cos rt$ series?" Can we express $f(t)$ as the following series:

$$y = \frac{A_0}{2} + \sum_{k=1}^{\infty}(A_k \cos kt + B_k \sin kt)$$

$$= \frac{A_0}{2} \; A_1 \cos t + B_1 \sin t + A_2 \cos 2t$$

$$+ B_2 \sin 2t + \ldots (.3)$$

where the function y is the displacement of a complex periodic motion of period $2n$ and is a sum of cosine and sine components with amplitudes A_1, A_2, ... B_1 B_2, ... and periods which are $2\pi/r$ (r being an integer). A_0 is a constant representing the displacement of the axis of motion from the t - axis.

The answer to the above question lies in the following theorem:

Fourier Theorem and Dirichlet Conditions: The theorem states that "any finite, continuous, single valued periodic function $f(t)$ of period 2π can be expressed as a summation of simple periodic terms having periods $2\pi/r$ where r is an integer".

Analytically Fourier theorem can be written as expansion of *f(t)* in Fourier series:

$$y \sim f\ (t) = \frac{A_0}{2} + A_1 \cos t + B_1 \sin t + A_2 \cos 2t + B_2 \sin 2t$$

$$+ A_3 \cos 3t + 3 \sin 3t + \qquad ...(3)$$

Alternatively if *f(x)* be defined in the interval (– *L, L)* and outside of this interval by *f(x + 2L) = f(x)* , i.e. assume that *f(x)* has the period L. The Fourier series corresponding to *f(x)* is given by

$$\frac{A_0}{2} + \sum_{n=1}^{\infty} A_n \cos\frac{nx\pi}{L} + B_n \sin\frac{n\pi x}{L} \qquad ...(4)$$

where the Fourier coefficient A_n and B_n are given by

$$A_n \frac{1}{L}\int_{-L}^{L} f(x)\cos\frac{\pi xn}{L}dx \qquad n = 0,1,2,3...$$

$$B_n = \frac{1}{L}\int_{-L}^{+L} f(x)\sin\frac{\pi xn}{L}dx \qquad ...(5)$$

As *f(x)* has a period 2L, the coefficient A_n, B_n can be determined equivalently from

$$A_n = \frac{1}{L}\int_a^{a+2L} f(x)\cos\frac{\pi xn}{L}dx$$

$$B_n = \frac{1}{L}\int_z^{a+2L} f(x)\sin\frac{\pi xn}{L}dx \qquad ...(6)$$

where a is a real number. For $a - L$ (6) reduces to (5)

In the case of *f(t)* in eq. (1) we have $L = \pi$ and the Coefficient are given by (5) or (6) by replacing $A_n \rightarrow a_n$ and $B_n \rightarrow b_n, L \rightarrow \pi$ It is to be noted that the expansion of a periodic function as Fourier series has some limitations. These limitations or conditions are known a Dirichlet Conditions:

Suppose that

(1) *f(x)* is defined and is single-valued except possibly at a finite number of points in (–L, L)

(2) $f(x)$ is periodic outside $(-L, L)$ with period 2L.

(3) $f(x)$ and $f(x) = \frac{df(x)}{dx}$ are piece-wise continuous in $(-L, L)$

then the series (2) with coefficients (3) or (4) converges to

(a) $f(x)$ if x is a point of continuity

(b) $\frac{f(x+0)+f(x\text{-}0)}{2}$ if x is a point of discontinuity

In the above $f(x + 0)$ and $f(x - 0)$ are the right and left limits of $f(x)$ at x. The conditions (1), (2), (3) imposed on $f(x)$ are sufficient by not necessary, and are generally satisfied in practice. These conditions are satisfied in many cases in physics and engineering , i.e. in the case of sound waves, plucked string problems, heat conduction, study state heat flow and numerous other problems. It is therefore necessary to study Fourier series separately and then apply them to various physical problems.

Orthogenality of sine and cosine functions:

(i) To prove eqs. (5) we integrate eq. (2) from –L to L to get

$$\frac{A_0}{2}\int_{-L}^{L} dx + \sum_{n=1}^{\infty}\int_{-L}^{L}\left(A_n \cos\frac{nx\pi}{L} + B_n \sin\frac{n\pi x}{L}\right)dx$$

$$= \frac{A_0}{2}(2L) + \sum_{n=1}^{\infty}\left\{A_n \int_{-L}^{L}\left(\cos\frac{n\pi x}{L}dx + B_n \int_{-L}^{L}\sin\frac{n\pi x}{L}dx\right)\right\}$$

$$= A_0 L + \sum_{n=1}^{\infty}\left\{A_n \left.\frac{\sin\frac{n\pi x}{L}}{\frac{n\pi}{L}}\right|_{-L}^{+L} - B_n \left.\frac{\cos\frac{n\pi x}{L}}{\frac{n\pi}{L}}\right|_{-L}^{+L}\right\}$$

$$= A_0 L + \sum_{n=1}^{\infty}\left[\frac{A_n L}{n\pi}\{\sin(n\pi) - \sin(-n\pi)\}\right.$$

$$\left. - \frac{B_n L}{n\pi}\{\cos(n\pi) - \cos(-n\pi)\}\right]$$

$$= A_0L + \sum_{n=1}^{\infty}\left[\frac{A_nL}{n\pi}(0) - \frac{B_nL}{n\pi}\{\cos n\pi - \cos n\pi\}\right]$$

$$= A_0L + \sum_{n=1}^{\infty}\left[0 - \frac{B_nL}{n\pi}(0)\right]$$

$$= A_0L$$

Thus, $A_0 = \frac{1}{L}\int_{-L}^{L} f(x)dx$

when the series (2) is equal to *f(x)*.

(ii) Similarly when we multiply eq. (2) by $\cos\frac{m\pi x}{L}$ and interact from –L to L, we get

$$\int_{-L}^{L} f(x)\cos\frac{m\pi x}{L}dx = \frac{A_0}{2}\int_{-L}^{L}\cos\frac{m\pi x}{L}dx$$

$$+\sum_{n=1}^{\infty}\left[A_n\int_{-L}^{L}\cos\frac{m\pi}{L}\cos\frac{m\pi x}{L}dx\right.$$

$$\left.+B_n\int_{-L}^{L}\cos\frac{m\pi x}{L}\sin\frac{m\pi x}{L}dx\right] \quad (8)$$

Now as before $\int_{-L}^{L}\cos\frac{m\pi x}{L}dx = 0$ for m ≠ 0 ...(8a)

and for m ≠ n

$$= \int_{-L}^{L}\cos\frac{m\pi x}{L}\cos\frac{m\pi x}{L}dx$$

$$= \frac{1}{2}\int_{-L}^{L}\left[\cos\frac{\pi(m+n)x}{L} + \cos\frac{\pi(m-n)x}{L}\right]dx$$

$$= \frac{1}{2}\left.\frac{\sin\frac{\pi(m+n)x}{L}}{\frac{\pi(m+n)}{L}}\right|_{-L}^{+L} + \frac{1}{2}\left.\frac{\sin\frac{\pi(m-n)x}{L}}{\frac{\pi(m+n)}{L}}\right|_{-L}^{+L}$$

$$= \frac{L}{2\pi(m+n)}\{\sin\pi(m+n) - \sin-\pi(m+n)\}$$

$$+\frac{L}{2\pi(m-n)}\{\sin\pi(m-n) - \sin[-\pi(m-n)]\}$$

$$= (0)+(0)=0$$

for m=n case

$$\int_{-L}^{L} \cos\frac{\pi mx}{L}\cos\frac{\pi nx}{L}dx = \int_{-L}^{L} \cos^2\frac{\pi mx}{L}dx$$

$$= \frac{1}{2}\int_{-L}^{L}\left(1+\cos\frac{2\pi mx}{L}\right)dx$$

$$= \frac{2L}{2}+0 = L$$

Thus, $\int_{-L}^{+L} \cos\frac{\pi mx}{L}\cos\frac{\pi nx}{L}dx = 0$ for $m \neq n$

$= L$ for $m \neq n$ ...(9)

In the same way for $m \neq$ n

$$\int_{-L}^{L} \cos\frac{\pi mx}{L}\sin\frac{\pi nx}{L}dx$$

$$= \frac{1}{2}\int_{-L}^{L}\left\{\sin\frac{\pi(n+m)x}{L} + \sin\frac{\pi(n-m)x}{L}\right\}dx$$

$$= \frac{1}{2}\left.\frac{\cos\frac{\pi(n+m)x}{L}}{\frac{\pi(n+m)}{L}}\right|_{-L}^{+L} + \frac{1}{2}\left.\frac{\cos\frac{\pi(n-m)x}{L}}{\frac{\pi(n-m)}{L}}\right|_{-L}^{+L}$$

and for $m = n$

$$\int_{-L}^{L} \cos\frac{\pi mx}{L}\sin\frac{\pi mx}{L}dx = \frac{1}{2}\int_{-L}^{L} \sin\frac{2\pi(m)x}{L}dx$$

$$= +\frac{1}{2}\left.\frac{\cos\frac{(2\pi mx)}{L}}{\frac{2\pi m}{L}}\right|_{-L}^{+L} = 0$$

Thus, $\int_{-L}^{L} \cos\frac{\pi mx}{L}\sin\frac{\pi mx}{L}dx = 0$ for all integer L and m. ...(10)

Thus, $\int_{-L}^{L} f(x)\cos\frac{m\pi x}{L}dx = \frac{A_0}{2}(0)$

$$+\sum_{n=1}^{\infty}\{A_n(L)\delta_{nm} + B_n(0)\}$$

or $\int_{-L}^{L} f(x)\cos\frac{m\pi x}{L}dx = LA_m$

or $A_m = \frac{1}{L}\int_{-L}^{L} f(x)\cos\frac{m\pi x}{L}dx$ ...(11)

(iii) When we multiply eq. (2) by $\sin\frac{m\pi x}{L}$ and interate from –L to L we get

$$\int_{-L}^{L} f(x)\sin\frac{m\pi x}{L}dx = \frac{A_0}{2}\int_{-L}^{L} \sin\frac{m\pi x}{L}dx$$

$$+\sum_{n=1}^{\infty}\left[A_n\int_{-L}^{L} \sin\frac{m\pi x}{L}\cos\frac{n\pi x}{L}dx + B_n\int_{-L}^{L} \sin\frac{m\pi x}{L}\sin\frac{n\pi x}{L}dx\right]$$

Again $\int_{-L}^{L}\sin\frac{m\pi x}{L}dx = 0$ for all m

Using eqs. (10), (12) we have

$$\int_{-L}^{L} f(x)\sin\frac{m\pi x}{L}dx = \sum_{n=1}^{\infty}B_n\int_{-L}^{L}\sin\frac{m\pi x}{L}\sin\frac{n\pi x}{L}dx$$

Now for $m \neq n$

$$\int_{-L}^{L}\sin\frac{m\pi x}{L}\sin\frac{n\pi x}{L}dx$$

$$= \frac{1}{2}\int_{-L}^{L}\left\{\cos\frac{(m-n)\pi x}{L} - \cos\frac{\pi x(m+n)}{L}\right\}dx$$

= (0) + (0) (for $m-n$ and $m+n$ as integers)

and for $m = n$

$$\int_{-L}^{L} \sin^2 \frac{m\pi x}{L} dx$$

$$= \frac{1}{2} \int_{-L}^{L} \left\{1 - \cos \frac{2m\pi x}{L}\right\} dx$$

$$= \frac{1}{2} \int_{-L}^{L} dx + 0 = L$$

Thus, $\int_{-L}^{L} \sin \frac{m\pi x}{L} \sin \frac{n\pi x}{L} dx = 0$ for $m \neq n$ ...(12)

= L for $m \neq n$

so $\int_{-L}^{L} f(x) \sin \frac{m\pi x}{L} dx = B_m L$

or $B_m = \frac{1}{L} \int_{-L}^{L} f(x) \sin \frac{m\pi x}{L} dx$ ...(13)

or

The relations, (8a) (9), (10), (12) and (13) are nothing but the orthogonality relations for sine and cosine functions.

Half Range Series

If a half-range series for a function $f(x)$ is desired, then the function is generally defined in the open interval $(0, \pi)$ [which is half of the interval$(-\pi, \pi)$ and hence the name half-range]. In this case the function $f(x)$ is expressed as an odd or even function; consequently the function is defined in the other half of the interval.

Let the series be

$$f(x) = \frac{a_0}{2} + \sum_{n=1}^{\infty} a_n \cos nx + \sum_{n=1}^{\infty} b_n \sin nx$$

In this case, if the function $f(x)$ is even, then

$$a_n = \frac{2}{\pi}\int_0^{\pi} f(x)\cos nx\, dx, \qquad \text{for } n = 0, 1, 2, 3, \ldots$$

and $\quad b_n = 0$ for $n = 1, 2, 3, \ldots$

and the function $f(x)$ is expressible in a series of cosines, i.e.,

$$f(x) = \frac{a_0}{2} + \sum_{n=1}^{\infty} a_n \cos nx \qquad \ldots(i)$$

But, if function $f(x)$ is odd, then

$a_n = 0$ for $n = 0, 1, 2, \ldots.$

and $$b_n = \frac{2}{\pi}\int_0^{\pi} f(x)\sin nx\, dx, \qquad \text{for } n = 1, 2, 3, \ldots$$

and $$f(x) = \sum_{n=1}^{\infty} b_n \sin nx$$

Significance of Cosine Series

If the function $f(x)$ is an even function of x, i. e., $f(-x) = f(x)$,

then $$\int_{-\pi}^{\pi} f(x)\sin nx\, dx = 0$$

which gives $b_n = 0$

and $$\frac{1}{\pi}\int_{-\pi}^{\pi} f(x)\cos nx\, dx = \frac{2}{\pi}\int_{-\pi}^{\pi} f(v)\cos nv\, dv = a_n \qquad \ldots(i)$$

Similarly,

$$a_0 = \frac{1}{2\pi}\int_{-\pi}^{\pi} f(v)\, dv = \frac{1}{\pi}\int_{-\pi}^{\pi} f(v)\, dv \qquad \ldots(ii)$$

Therefore, the Fourier's expansion reduces to

$$f(x) = a_0 + \sum_{n=1}^{\infty} a_n \cos nx \qquad \ldots(iii)$$

$$= \frac{1}{\pi} \int_{-\pi}^{\pi} f(v)\, dv + \frac{2}{\pi} \sum_{n=1}^{x} \left[\int_{0}^{\pi} f(v) \cos nv\, dv \right] \cos nx \qquad \text{...(iv)}$$

which represents the function *f(x)* in a series of cosines and therefore, eq. (iv) is known as *Cosine series* in the interval (0, π)

Value of Sine Series

When the function *f(x)* is an odd function of *x, i.e., f(x),* we have

$$\int_{-\pi}^{\pi} f(v)\, dv = 0$$

which gives $a_0 = 0$,

giving $a_n = 0$

$$\text{Also } \frac{1}{\pi} \int_{-\pi}^{\pi} f(v) \sin nv\, dv = \frac{2}{\pi} \int_{0}^{\pi} f(v) \sin nv\, dv = b_n \qquad \text{...(v)}$$

Substituting these values, we get

$$f(x) = \sum_{n=1}^{\infty} b_n \sin nx$$

$$= \frac{2}{\pi} \sum_{n=1}^{\infty} \sin nx \int_{0}^{\pi} f(v) \sin nv\, dv \qquad \text{...(vi)}$$

which represents the function *f(x)* in a series of sines in the interval of (0, π). It is known as *sine series.*

Change of Interval

Suppose we want to represent the function *f(x)* defined in the closed interval (–c, c) by a Fourier series, *c* being any positive real number. We consider this interval as a result of elongating (or compressing) the interval [–π, π]. The interval $-c \leq x \leq c$ is transformed into the interval $-\pi \leq z \leq \pi$ by the transformation.

$$z = \pi \frac{x}{c}$$

Then $f(x) = f\left(\frac{cz}{\pi}\right) = F(z)$ (*say*).

Let $F(z) = \frac{a_0}{2} + \sum_{n=1}^{\infty}(a_n \cos nz + b_n \sin nz)$

with $a_{n=} \frac{1}{\pi}\int_{-\pi}^{\pi} F(z)\cos nz\, dz,$ (n=0,1,2,3,...)

$$b_n = \frac{1}{\pi}\int_{-\pi}^{\pi} F(z)\sin nz\, dz\ (n = 0, 1, 2, 3, ...)$$

Applying the transformation $z = \pi \frac{x}{c}$ so that

$dz = \frac{\pi}{c}dx$, we get

$$f(x) = \frac{a_0}{2} + \sum_{n=1}^{\infty} a_n \cos\left(\frac{n\pi x}{c}\right) + d_n \sin\left(\frac{n\pi x}{c}\right)$$

with $a_n = \frac{1}{c}\int_{-c}^{c} f(x)\cos\frac{n\pi x}{c}dx, (n = 0, 1, 2, 3, ...)$

$$b_n = \frac{1}{c}\int_{-c}^{c} f(x)\sin\frac{n\pi x}{c}dx, (n = 1, 2, 3, ...)$$

Change of Period

By means of the formula

$$z = Ax = B \qquad ...(i)$$

we can change any interval $[a, b]$ to $[-\pi, \pi]$.

Here we have $x = a$ when $z = -\pi$, and $x = b$ when $z = \pi$.

$\therefore \quad -\pi = Aa + B, \pi = Ab + B.$

From which, we get

$$A = \frac{2\pi}{b-a}, B = -\frac{\pi(a+b)}{b-a}$$

Thus, the transformation

$$z = \frac{2\pi x}{b-a} + \frac{\pi(a+b)}{b-a} = \frac{\pi}{b-a}(2x-a-b)$$

transforms the interval $a \leq x \leq b$ into $-\pi \leq z \leq \pi$.

Fourier Series: Complex Form and Definition

In complex notation, the Fourier series is written as

$$f(x) = \sum_{n=\infty}^{\infty} c_n . e^{inx}, (n = 0, 1, 2, ...)$$

or equivalently,

$$f(x) = c_0 + \sum_{n=1}^{\infty} c_n e^{inx} + \sum_{n=1}^{\infty} c_{-n} e^{inx} \quad ...(i)$$

with $$c_{-n} = \frac{1}{2\pi}\int_c^{c+2\pi} f(x) e^{-inx} dx, \qquad (n = 0, 1, 2, ...);$$

$$c_{-n} = \frac{1}{2\pi}\int_c^{c+2\pi} f(x) e^{inx} dx, \qquad (n = 0, 1, 2, ...);$$

f (x) being defined in the interval [c, c + 2 π]

Here $$c_0 = \frac{1}{2\pi}\int_c^{c+2\pi} f(x)\, dx = \frac{a_0}{2}$$

$$c_n = \frac{1}{2\pi}\int_c^{c+2\pi} f(x)\, e^{-inx}\, dx$$

$$= \frac{1}{2\pi}\int_c^{c+2\pi} f(x)\, (\cos nx - \sin nx) dx$$

$$= \frac{1}{2\pi}\int_c^{c+2\pi} f(x) \cos nx - i.\frac{1}{2\pi}\int_c^{c=2\pi} f(x) \sin nx\, dx$$

$$= \frac{a_{n-ib_n}}{2}$$

Similarly,

$$c_{-n} = \frac{1}{2\pi} \int_{c}^{c+2\pi} f(x)\, e^{-inx}\, dx$$

gives $$c_{-n} = \frac{a_n - ib_n}{2}$$

Now (i) becomes

$$f(x) = \frac{a_0}{2} + \sum_{n=1}^{\infty} \left\{ \frac{(a_n - ib_n)}{2} \right\} \cos nx$$

$$+ \sum_{n=1}^{\infty} \left\{ \frac{(a_n + ib_n)}{2} \right\} \sin nx \qquad ...(ii)$$

with $$a_n = \frac{1}{\pi} \int_{c}^{c+2\pi} f(x) \sin nx\, dx, \quad (n=0,1,2,3,...)$$

$$b_n = \frac{1}{\pi} \int_{c}^{c+2\pi} f(x) \sin nx\, dx, \quad (n=0,1,2,3,...)$$

(ii) is an alternate form of (i).

Properties of Fourier Series (Distinctive Features)

(1) *Convergence:* It is to be noted that the Fourier series representing a discontinuous function will not be uniformly convergent. A uniformly convergent series of continuous function always yields a continuous function. The Fourier series of a function *f(x)* will converge uniformly if the function *f(x)* satisfies following conditions:

(a) *f(x)* is continuous in the interval $-\pi \leq x \leq \pi$

(b) $f(-\pi) = f(+\pi)$

(c) *f(x)* is sectionally continuous.

These restrictions do not demand that $f(x)$ be periodic but they are satisfied by continuous, differentiable, periodic function (period = 2π).

(ii) *Interaction:* The term by term integration of Fourier series yields

$$\int_{x_0}^{x} f(x)\,dx = [a_0 x]_{x0}^{x}$$

$$+\left[\sum_{n=1}^{\infty}\frac{a_n}{n}\sin nx\right]_{x0}^{x} - \left[\sum_{n=1}^{\infty}\frac{b_n \cos nx}{n}\right]_{x_0}^{x}$$

Thus, the effect of integration is to place an additional power of n in the denominator of each coefficient. This result is more rapidly convergent.

Hence, convergent Fourier series may always be integrated term by term and the resulting series converges uniformly to the integral of the original function.

The series given by equation (i) is not a Fourier series if $a_0 \neq 0$ but

$$\int_{x_0}^{x} f(x)\,dx - a_0 x$$

will still be a Fourier series.

(iii) Differentiation: The operation of differentiation places an additional factor n in the numerator of each term of Fourier series. This reduces the rate of convergence and the differentiated series may be divergent also.

(iv) Completeness Relation (Parseval's Theorem): One of the important properties of Fourier series is completeness relation of Parseval's theorem, which gives a relation between the average of the square (or absolute square)

of the function *f(x)* and the coefficients in Fourier series for *f(x)*

Let the Fourier's expansion be

$$f(x) = a_0 + \sum_{n=1}^{\infty} a_n \cos nx + \sum_{n=1}^{\infty} b_n \sin nx$$

Average of the square of f(x) over $[-\pi, \pi]$ is

$$\frac{1}{2\pi}\int_{-\pi}^{\pi}[f(x)]^2\, dx.$$

Thus, we have the average of

$$[f(x)]^2 = \frac{1}{2\pi}\int_{-\pi}^{\pi}\left[a_0 + \sum_{n=1}^{\infty} a_n \cos nx + \sum_{n=1}^{\infty} b_b \sin nx\right]^2 dx$$

$$= \frac{1}{2\pi}\int_{-\pi}^{\pi}(a_0)^2 dx + \frac{1}{2\pi}\sum_{n=1}^{\infty}\int_{-\pi}^{\pi} a_{n^2} \cos^2 nx\, dx$$

$$+ \frac{1}{2\pi}\sum_{n=1}^{\infty}\int_{-\pi}^{\pi} b_n^2 \sin^2 nx\, dx \qquad \text{...(ii)}$$

other terms vanish when average is taken. Evaluating the interals in this equation, we get average of $[f(x)]^2$ (over the given period)

$$= a_0^2 + \frac{1}{2}\sum_{n=1}^{\infty} a_n^2 + \frac{1}{2}\sum_{n=1}^{\infty} b_n^2 \qquad \text{...(iii)}$$

This is one form of Parseval theorem. It can be easily verified that the theorem is unchanged if *f(x)* has period 2*l* instead of 2π and its square is averaged over any period of length 2*l*. It can also be verified that if *f(x)* be written as complex exponential Fourier series, then we find average of

$$|f(x)|^2 = \sum_{-\infty}^{\infty} |c_n|^2$$

ILLUSTRATIVE EXAMPLES

1. Expand in a Fourier series the following function f(x) defined in the interval $-\pi < x < \pi$

$$f(x) = x \text{ when } -\pi < x < \pi$$

Solution: Let the given function be expanded in a Fourier series

$$f(x) = A_0 + \sum_{r=1}^{\infty}(A_r \cos rx + B_r \sin rx),$$

where the coefficients A_0, A_r and B_r are to be evaluated

$$A_0 = \frac{1}{2\pi}\int_{-\pi}^{\pi} f(x)dx$$

$$= \frac{1}{2\pi}\int_{-\pi}^{\pi} xdx$$

$$= \frac{1}{2\pi}\left[\frac{x^2}{2}\right]_{-\pi}^{\pi} = \frac{1}{2\pi}\left[\frac{\pi^2}{2} - \frac{(-\pi)^2}{2}\right] = 0.$$

Again,

$$A_r = \frac{1}{\pi}\int_{-\pi}^{\pi} f(x)\cos rx\, dx \qquad r = 1, 2, \ldots$$

$$= \frac{1}{\pi}\int_{-\pi}^{\pi} x\cos rx\, dx$$

$$= \frac{1}{\pi}\left[x\frac{\sin rx}{r} - \int\frac{\sin rx}{r}dx\right]_{-\pi}^{\pi} \qquad \text{(by parts)}$$

$$= \frac{1}{\pi}\left[x\frac{\sin rx}{r} + \frac{\sin rx}{r^2}\right]_{-\pi}^{\pi}$$

$$= \frac{1}{\pi r^2}[rx\sin rx + \cos rx]_{-\pi}^{\pi}$$

$$= \frac{1}{\pi r^2}[\cos rx - \cos(-rx)] \qquad [\because \sin rx = 0 \text{ at } \pi, -\pi]$$

$$l = \frac{1}{\pi r^2}[\cos rx - \cos rx] \qquad [\because \cos(-rx) = \cos rx]$$

Similarly

$$B_r = \frac{1}{\pi}\int_{-\pi}^{\pi} f(x)\sin rx\, dx \qquad r=1,\ 2,\ \ldots$$

$$= \frac{1}{\pi}\int_{-\pi}^{\pi} x\sin rx\, dx$$

$$= \frac{1}{\pi} x\left(-\frac{\cos rx}{r}\right) - \int -\frac{\cos rx}{r}dx\Bigg]_{-\pi}^{\pi}$$

$$= \frac{1}{\pi}\left(-x\frac{\cos rx}{r}\right) + \frac{\sin rx}{r^2}\Bigg]_{-\pi}^{\pi}$$

$$= \frac{1}{\pi r^2}[-rx\cos rx + \sin\ rx]_{-\pi}^{\pi}$$

$$= \frac{1}{\pi r^2}[-rx\cos rx + r(-\pi)\cos(-rx)]$$

$$= \frac{1}{\pi r^2}[-rx\cos rx - rx\cos rx]$$

$$= \frac{1}{\pi r^2}[-2r\pi\cos r\pi]$$

$$= -\frac{2}{r}\cos r\pi$$

$$= -\frac{2}{r}\begin{Bmatrix}1 \text{ for even } r\\ -1 \text{ for odd } r\end{Bmatrix}$$

$$= -\frac{2}{r}(-1)^r$$

Substituting the values of A_0 A_r and B_r in the Fourier series,

we get

$$f(x) = \sum_{r=1}^{\infty} -\frac{2}{r}(-1)^r \sin rx$$

$$= 2\sum_{r=1}^{\infty} -\frac{(-1)^{r+1}}{r}\sin rx$$

$$= 2\left(\sin x - \frac{1}{2}\sin 2x + \frac{1}{3}\sin 3x - \frac{1}{4}\sin 4x + \ldots\right)$$

2. Write the Fourier series corresponding to the function f (x) defined in the interval – π< x < π as follows:

$$f(x) = 0 \text{ when } -\pi < x < 0$$

$$= x \text{ when } 0 < x < \pi.$$

Solution: The given function may be written in a Fourier series as:

$$f(x) = A_0 + \sum_{r=1}^{\infty}(A_r \cos rx = B_r \sin rs),$$

where A_0, A_r and B_r are the Fourier coefficients. These are to be evaluated.

$$A_0 = \frac{1}{2\pi}\int_{-\pi}^{\pi} fx\,dx$$

$$= \frac{1}{2\pi}\left[\int_{-n}^{0} 0 + \int_{0}^{\pi}(x)\,dx\right]$$

$$= \frac{1}{2\pi}\left[\frac{x^2}{2}\right]_0^{\pi} = \frac{\pi}{4}$$

Again,

$$A_r = \frac{1}{\pi}\int_{-\pi}^{\pi} f(x)\cos rx\,dx, \qquad r = 1, 2, ...$$

$$= \frac{1}{\pi}\int_0^{\pi} x\cos rs\,dx$$

$$= \frac{1}{\pi}\left[x\frac{\sin rx}{r} - \int\frac{\sin rx}{r}dx\right]_0^{\pi} \qquad \text{(by parts)}$$

$$= \frac{1}{\pi}\left[x\frac{\sin rx}{r} + \int\frac{\cos rx}{r^2}\right]_0^{\pi}$$

$$= \frac{1}{\pi r^2}[rx\sin rx + \cos rx]_0^{\pi}$$

$$= \frac{1}{\pi r^2}[\cos r\pi - 1]$$

$$= \frac{1}{\pi r^2}\left[\begin{Bmatrix} 1 \text{ for even } r \\ -1 \text{ for odd } r \end{Bmatrix} - 1\right]$$

$$= \frac{1}{\pi r^2}\left[(-1)^r - 1\right].$$

Similarly ,

$$B_r = \frac{1}{\pi}\int_{-\pi}^{\pi} f(x)\sin rx\, dx, \qquad r = 1,2,...$$

$$= \frac{1}{\pi}\int_0^{\pi} x\sin rx\, dx$$

$$= \frac{1}{\pi}\left[x\left(-\frac{\cos rx}{r} - \int\frac{\cos rx}{r}dx\right)\right]_0^{\pi} \qquad \text{(by parts)}$$

$$= \frac{1}{\pi}\left[-x\frac{\cos rx}{r} + \frac{\sin rx}{r^2}\right]_0^{\pi}$$

$$= \frac{1}{\pi r^2}\left[-rx\cos rs + \sin rx\right]^{\pi}$$

$$= \frac{1}{\pi r^2}\left[-r\pi\cos r\pi\right]$$

$$= -\frac{1}{r}\cos r\pi$$

$$= -\frac{1}{r}\begin{Bmatrix} 1 \text{ for even } r \\ \text{-1 for odd } r \end{Bmatrix}$$

$$= -\frac{1}{r}(-1)^r$$

$$= -\frac{1}{r}(-1)^{r+1}$$

Substituting the values of A_0, A_r and B_r in the Fourier series, we get

$$f(x) = \frac{\pi}{4} + \sum_{r=1}^{\infty}\left[\frac{(-1)^r - 1}{\pi r^2}\cos rx + \frac{(-1)r+1}{r}\sin rx\right], \quad r = 1,2,3,....$$

$$= \frac{\pi}{4} + \left(-\frac{2}{\pi} \cos x + \sin x \right) - \frac{1}{2} \sin 2x$$

$$+ \left(\frac{2}{9\pi} \cos 3x + \frac{1}{3} \sin 3x \right) - \frac{1}{4} \sin 4x + ...$$

3. *Find the Fourier series of the following function which is assumed to have the period* 2π

$$f(x) = \begin{Bmatrix} -x \text{ for } -\pi < x < 0 \\ x \text{ for } 0 < x < \pi \end{Bmatrix}.$$

Solution. Let us write

$$f(x) = A_0 + \sum_{r=1}^{\infty} (A_r \cos rx + B_r \sin rx),$$

and evaluate the fourier coefficients A_0, A_r and B_r .

$$A_0 = \frac{1}{2\pi} \int_{-\pi}^{\pi} f(x) dx$$

$$= \frac{1}{2\pi} \left[- \int_{-\pi}^{0} x \, dx + \int_{0}^{\pi} x \, dx \right]$$

$$= \frac{1}{2\pi} \left[\frac{\pi^2}{2} + \frac{\pi^2}{2} \right] = \frac{\pi}{2}.$$

$$A_r = \frac{1}{\pi} \int_{-\pi}^{\pi} f(x) \cos rx \, dx, \quad r = 1, 2, ...$$

$$= \frac{1}{\pi} \left[- \int_{-\pi}^{0} x \cos rx \, dx + \int_{0}^{\pi} x \cos rx \, dx \right]$$

$$= \frac{1}{\pi r^2} \left[-\{rx \sin rs + \cos rx\}_{-\pi}^{0} + \{rx \sin + \cos rx\}_{-\pi}^{0} \right]$$

$$= \frac{1}{\pi r^2} [-\{1 - \cos rx\} + \{\cos rx - 1\}]$$

$$=\frac{2}{\pi r^2}[\cos r\pi - 1]$$

$$=\frac{2}{\pi r^2}\left[(-1)^r - 1\right]$$

$$B_r = \frac{1}{\pi}\int_{=\pi}^{\pi} f(x)\, si\, rx\, dx, \qquad r = 1, 2, 3,...$$

$$=\frac{1}{\pi}\left[-\{-rx\cos rx + \sin rx\}_{-\pi}^{0} + \{-rx\cos rx + \sin rx\}_{0}^{\pi}\right]$$

$$=\frac{1}{\pi r^2}\left[-\{-rx\cos rx + \sin rx\}_{-\pi}^{0} + \{-rx\cos rx + \sin rx\}_{0}^{\pi}\right]$$

$$=\frac{1}{\pi r^2}\left[-\{-r\pi\cos r\pi\} + \{-r\pi\cos r\pi\}\right]$$

Substituting the value of A_0, A_r and B_r in the Fourier series, we get

$$f(x) = \frac{\pi}{2} + \sum_{r=1}^{\infty} \frac{2}{\pi r2}\left[(-1)^r - 1\right]\cos rx$$

$$=\frac{\pi}{2} - \frac{4}{\pi}\left(\cos x + \frac{1}{9}\cos 3x + \frac{1}{25}\cos 5x + ...\right)$$

4. *Expand in a Fourier series*

$$f(x) = \begin{cases} 0(-\pi \le x \le 0) \\ h(0 \le x \le \pi) \end{cases}.$$

Solution. Let us write

$$f(x) = A_0 + \sum_{r=1}^{\infty}\left(A_r \cos rx + B_r \sin rx\right),$$

where $A_0 = \frac{1}{2\pi}\int_{-\pi}^{\pi} f(x)dx, A_r = \frac{1}{\pi}\int_{-\pi}^{\pi} f(x)\cos rx\, dx$

and $B_r = \frac{1}{\pi}\int_{-\pi}^{\pi} f(x)\sin rx\, dx.$ Let us evaluate A_0, A_r, *and* B_r

$$A_0 = \frac{1}{2\pi}\int_{-\pi}^{\pi} f(x)\,dx,$$

$$= \frac{1}{2\pi}\left[\int_{-\pi}^{0} 0 + \int_{0}^{\pi} h\,dx\right]$$

$$= \frac{h}{2\pi}[x]_0^{\pi} = \frac{h}{2}$$

$$A_r = \frac{1}{\pi}\int_{-\pi}^{\pi} f(x)\cos rx\,dx$$

$$= \frac{1}{\pi}\int_{0}^{\pi} h\cos rx\,dx$$

$$= \frac{h}{\pi}\left[\frac{\sin rx}{r}\right]_0^{\pi} = 0$$

$$B_r = \frac{1}{\pi}\int_{-\pi}^{\pi} f(x)\sin rx\,dx$$

$$= \frac{1}{\pi}\int_{0}^{\pi} h\sin rx\,dx$$

$$= \frac{h}{\pi}\left[-\frac{\cos rx}{r}\right]_0^{\pi}$$

$$= \frac{h}{r\pi}[-\cos r\pi + \cos 0]$$

$$= \frac{h}{r\pi}[1 - \cos\ r\pi]$$

$$= \frac{h}{r\pi}\left[1-(-1)^r\right] \quad [\because \cos r\pi \text{ is -1 for odd } r \text{ and 1 for even r.}]$$

Substituting the values of A_0, A_r and B_r in the Fourier's series, we get

$$f(x) = \frac{h}{2} + \sum_{r=1}^{\infty}\frac{h}{r\pi}\left[1-(1)^r\right]\sin rx$$

$$= \frac{h}{2} + \frac{2h}{\pi}\left(\sin x + \frac{\sin 3x}{3} + \frac{\sin 5x}{5} + \ldots\right)$$

5. *A square wave is represented by the following function:*

$$f(x) = h \text{ for } 0 < x < \pi$$

and $$f(x) = -h \text{ for } \pi < x < 2\pi.$$

Find the Fourier series representation of the wave.

Solution. The Fourier series expansion of a function *f(x)* is written as:

$$F(x) = A_0 + \sum_{r=1}^{\infty}(A_r \cos rx + B_r \sin rx)$$

where A_0, A_r and B_r are Fourier coefficients.

$$A_0 = \frac{1}{2\pi}\int_{-\pi}^{\pi} f(x)\,dx$$

$$= \frac{1}{2\pi}\left[\int_0^{\pi} h\,dx - \int_{\pi}^{2\pi} h\,dx\right]$$

$$= \frac{h}{2\pi}\left[|x|_0^{\pi} - |x|_{\pi}^{2\pi}\right] = 0.$$

$$A_r = \frac{1}{\pi}\int_{-\pi}^{\pi} f(x)\cos rx\,dx, \qquad r = 1, 2, 3, \ldots$$

$$= \frac{1}{\pi}\left[\int_0^{\pi} h\cos rd\,dx - \int_{\pi}^{2\pi} h\cos rx\,ds\right]$$

$$= \frac{h}{\pi r}\left[|\sin rx|_0^{\pi} - |\sin rx|_{\pi}^{2\pi}\right] = 0$$

$$B_r = \frac{1}{\pi}\int_{-\pi}^{\pi} f(x)\sin rx\,dx$$

$$= \frac{1}{\pi}\left[\int_0^{\pi} h\sin rx\,dx - \int_{\pi}^{2\pi} h\sin rd\,dx\right]$$

$$= \frac{h}{\pi r}\left[\,|-\cos rx|_0^{\pi} - |-\cos rx|_{\pi}^{2\pi}\right]$$

$$= \frac{h}{\pi r}\left[\left(\cos r\pi + \cos 0\right) + \left(\cos 2r\pi - \cos r\pi\right)\right]$$

$$= \frac{h}{\pi r}\left[-\cos r\pi + 1 + 1 - \cos r\pi\right]$$

$$= \frac{2h}{\pi r}\left[1 - \cos r\pi\right]$$

$$= \frac{2h}{\pi r}\left[1 - (-1)^r\right]$$

Substituting the values of A_0, A_r and B_r in the Fourier series expansion, we get

$$f(x) = \sum_{r=1}^{\infty} \frac{2h}{r\pi}\left[1 - (-1)^r\right] \sin rx$$

$$= \frac{4h}{\pi}\left(\sin x + \frac{\sin 3x}{3} + \frac{\sin 5x}{5} + \ldots\right)$$

6. Find the Fourier series of the function

$f(x) = x + \pi$ when $-\pi < x < \pi$ and $f(x + 2\pi) = f(x)$.

Solution. The given function can be expanded in a Fourier series:

$$f(x) = A_0 + \sum_{r=1}^{\infty}\left(A_r \cos rx + B_r \sin rx\right),$$

Where the coefficient A_0, A_r and B_r are to be evaluated.

$$A_0 = \frac{1}{2\pi}\int_{-\pi}^{\pi} f(x)\,dx$$

$$= \frac{1}{2\pi}\int_{-\pi}^{\pi} (x + \pi)\,dx$$

$$= \frac{1}{2\pi}\left[\frac{x^2}{2} + \pi x\right]_{-\pi}^{\pi}$$

$$= \frac{1}{2\pi}\left[\left(\frac{\pi^2}{2} + \pi^2\right) - \frac{\pi^2}{2} - \pi^2\right]$$

$$= \pi$$

$$A_r = \frac{1}{\pi}\int_{-\pi}^{\pi} f(x)\cos rx\, dx, \qquad r = 1,2,3,...$$

$$= \frac{1}{\pi}\int_{-\pi}^{\pi}(x+\pi)\cos rx\, dx$$

$$= \frac{1}{\pi}\left[\int_{-\pi}^{\pi} x\cos rx\, dx + \pi\int_{-\pi}^{\pi}\cos rx\, dx\right]$$

$$= \frac{1}{\pi}\left[\left\{x\frac{\sin rx}{r} + \frac{\cos r\pi}{r_2}\right\}_{-\pi}^{\pi} + \pi\left\{\frac{\sin rx}{2}\right\}_{-\pi}^{\pi}\right]$$

$$= \frac{1}{\pi}\left[\left\{0 + \frac{\cos r\pi}{r^2} - 0\frac{\cos(-r\pi)}{r^2}\right\} + \pi\{0\}\right] = 0$$

$$B_r = \frac{1}{\pi}\int_{-\pi}^{\pi} f(x)\sin rx\, dx, \qquad r = 1,2,3,...$$

$$= \frac{1}{\pi}\left[\int_{-\pi}^{\pi} x\sin rx\, dx = \pi\int_{-\pi}^{\pi}\sin rx\, dx\right]$$

$$= \frac{1}{\pi}\left[\left\{-x\frac{\cos rx}{r} + \frac{\sin rx}{r^2}\right\} - \pi\left\{\frac{\cos rx}{r}\right\}_{\pi}^{\pi}\right]$$

$$= \frac{1}{\pi r^2}\{-rx\cos rx + \sin rx\}_{-\pi}^{\pi} \quad -\frac{1}{r}\{\cos rx\}_{-\pi}^{\pi}$$

$$= \frac{1}{\pi r^2}\{-r\pi\cos r\pi + r(-\pi)\cos(-r\pi)\} - \frac{1}{r}\{\cos r\pi - \cos(-r\pi)\}$$

$$= -\frac{2}{r}\cos r\pi$$

$$= -\frac{2}{r}(-1)^r$$

Substituting the values of A_0, A_r and B_r in the Fourier series expansion, we get

$$f(x) = \pi + \sum_{r=1}^{\infty} -\frac{2}{r}(-1)^r \sin rx$$

$$= \pi + \sum_{r=1}^{\infty} -\frac{(-1)^{r-1}}{r}\sin rx$$

$$= \pi + 2\left(\sin x - \frac{1}{2}\sin 2x + \frac{1}{3}\sin 3x - \frac{1}{4}\sin 4x + ...\right)$$

7. Expand in Fourier series

$$f(x) = \begin{cases} 0 \text{ when } -\pi < x < 0 \\ \sin x \text{ when } 0 < x < \pi. \end{cases}$$

Solution: Let us write *f(x)* as a Fourier series:

$$f(x) = A_0 + \sum_{r=1}^{\infty} (A_r \cos rx + B_r \sin rx),$$

and evaluate the coefficients A_0, A_r and B_r.

$$A_0 = \frac{1}{2\pi} \int_{-\pi}^{\pi} f(x)\, dx$$

$$= \frac{1}{2\pi} \left[\int_{-\pi}^{0} 0 dx + \int_{0}^{\pi} \sin x\, dx \right]$$

$$= \frac{1}{2\pi} [-\cos x]_0^{\pi} = \frac{1}{\pi}$$

$$A_r = \frac{1}{\pi} \int_{-\pi}^{\pi} f(x) \cos rx\, dx, \qquad r = 1, 2, \ldots$$

$$= \frac{1}{\pi} \int_{0}^{\pi} \sin x \cos rx\, dx$$

$$= \frac{1}{2\pi} \int_{0}^{\pi} \{\sin(x + rx) + \sin(x - rx)\}\, dx$$

$$= \frac{1}{2\pi} \left[-\frac{\cos(1+r)x}{1+r} - \frac{\cos(1-r)x}{1-r} \right]_0^{\pi}$$

$$= \frac{1}{2\pi} \left[-\frac{\cos(1+r)\pi - 1}{1-r} - \frac{\cos(1-r)\pi - 1}{1-r} \right]$$

$$= \frac{1}{2\pi} \left[-\frac{1-\cos(1+r)\pi}{1+r} + \frac{1-\cos(1-r)\pi}{1-r} \right]$$

$$= \frac{1}{2\pi} \left[\frac{1+\cos r\pi}{1+r} + \frac{1+\cos r\pi}{1-r} \right]$$

$$= \frac{1}{2\pi} \left[(1+\cos r\pi) \left[\frac{1}{1+r} + \frac{1}{1-r} \right] \right]$$

$$= \frac{1}{\pi}\frac{1+r\pi}{1-r^2}$$

$$= \frac{-(1+\cos r\pi)}{\pi(r^2-1)} \qquad \text{if } r \neq 1$$

For r = 1,

$$A_1 = \frac{1}{\pi}\int_0^{\pi} \sin x \cos x \, dx$$

$$= \frac{1}{\pi}\int_0^{\pi} \frac{\sin 2x}{2} dx \frac{1}{\pi}\left[-\frac{\cos 2x}{4}\right]_0^{\pi} = 0$$

$$B_r = \frac{1}{\pi}\int_{-\pi}^{\pi} f(x) \sin rx \, dx \qquad r = 1, 2, \ldots$$

$$= \frac{1}{\pi}\int_0^{\pi} \sin x \sin rx \, dx$$

$$= \frac{1}{2\pi}\int_0^{\pi} \{\cos(x - rx) - \cos(x + rx)\} dx$$

$$= \frac{1}{2\pi}\left[\frac{\sin(1-r)x}{1-r} - \frac{\sin(1-r)x}{1-r}\right]_0^{\pi}$$

$$= \frac{1}{2\pi}\left[\frac{\sin(1-r)\pi}{1-r} + \frac{\sin(1+r)\pi}{1+r}\right]$$

$$= \frac{1}{2\pi}(\sin r\pi)\left[\frac{1}{1-r} + \frac{1}{1+r}\right]$$

$$= \frac{1}{\pi}\frac{\sin r\pi}{1-r^2}$$

$$= 0 \text{ if } r \neq 1$$

For $r = 1$, $B_1 = \frac{1}{\pi}\int_0^{\pi} \sin^2 x dx = \frac{1}{\pi}\int_0^{\pi} \frac{1-\cos 2x}{2} dx$

$$= \frac{1}{2\pi}\left[x - \frac{\sin 2x}{2}\right]_0^{\pi} = \frac{1}{2\pi}[\pi] = \frac{1}{2}$$

Substituting the values of A_0, A_r and B_r in Fourier series, we have

$$f(x) = \frac{1}{\pi} + \sum_{r=2}^{\infty}\left[\frac{-(1+\cos r\pi)}{\pi(r^2-1)}\cos rx\right] + \frac{1}{2}\sin x$$

$$\left[\because B_1 = \frac{1}{2}, B_2 = B_3 = ...01\right]$$

$$= \frac{1}{\pi} + \frac{1}{2}\sin x - \frac{1}{\pi}\sum_{r=2}^{\infty}\frac{1+\cos r\pi}{r^2-1}\cos rx$$

$$= \frac{1}{\pi} + \frac{1}{2}\sin x - \frac{2}{\pi}\left(\frac{\cos 2x}{2^3-1} + \frac{\cos 4x}{4^2-1} + \frac{\cos 6x}{6^2-1} + ...\right).$$

8. Expand in a Fourier series:

$$f(x) = \begin{bmatrix} \sin x(-0 < x < \pi) \\ -\sin x(-\pi < x < 0) \end{bmatrix}$$

Solution: Let us write *f(x)* as a Fourier series:

$$f(x) = A_0 + \sum_{r=1}^{\infty}(A_r \cos rx + B_r \sin rx)$$

and evaluate the coefficients A_0, A_r and B_r

$$A_0 = \frac{1}{2\pi}\int_{-\pi}^{\pi} f(x)dx$$

$$= \frac{1}{2\pi}\left[\int_0^{\pi}\sin x\, dx - \int_{-\pi}^{0}\sin x\, dx\right]$$

$$= \frac{1}{2\pi}\left[\{-\cos x\}_0^{\pi} - \{-\cos x\}_{-\pi}^{0}\right]$$

$$= \frac{1}{2\pi}\left[\{-\cos\pi + \cos 0\} - \{-\cos 0 + \cos(-\pi)\}\right]$$

$$= \frac{1}{2\pi}[[1+1]-[-1-1]] = \frac{2}{\pi}$$

$$A_r = \frac{1}{\pi}\int_{-\pi}^{\pi} f(x)\cos rx\,dx,$$

$$= \frac{1}{\pi}\left[\int_0^{\pi}\sin x\cos rx\,dx - \int_{-\pi}^{0}\sin x\cos rx\,dx\right]$$

$$= \frac{-2(1+\cos r\pi)}{\pi(r^2-1)} \text{ if } r \neq 1 \qquad \text{(Solve as in last prob.)}$$

For $r = 1$, $$A_1 = \frac{1}{\pi}\left[\int_0^{\pi}\sin x\cos x\,dx - \int_{-\pi}^{0}\sin x\cos x\,dx\right] = 0$$

$$B_r = \frac{1}{\pi}\int_{-\pi}^{\pi} f(x)\sin rx\,dx, \qquad r = 1, 2, \ldots$$

$$= \frac{1}{\pi}\left[\int_0^{\pi}\sin x\sin rx\,dx - \int_{-\pi}^{0}\sin x\sin rx\,dx\right]$$

$$= 0, \text{if } r \neq 1 \qquad \text{(Solve as in last prob.)}$$

For $r = 1$, $$B_1 = \frac{1}{\pi}\left[\int_0^{\pi}\sin^2 x\,dx - \int_{-\pi}^{0}\sin^2 x\,dx\right]$$

$$= \frac{1}{\pi}\left[\int_0^{\pi}(1-\cos 2x)\,dx - \int_{-\pi}^{0}(1-\cos 2x)\,dx\right]$$

$$= \frac{1}{2\pi}\left[\left\{x - \frac{\sin 2x}{2}\right\}_0^{\pi} - \left\{x - \frac{\sin 2x}{2}\right\}_{-\pi}^{0}\right]$$

$$= \frac{1}{2\pi}\left[\{\pi\} - \{\pi\}\right] = 0.$$

Substituting the values of A_0, A_r and B_r in Fourier series, we have

$$f(x) = \frac{2}{\pi} + \sum_{r=2}^{\infty}\frac{-2(1+\cos r\pi)}{\pi(r^2-1)}\cos rx$$

$$= \frac{2}{\pi} - \frac{2}{\pi}\sum_{r=2}^{\infty}\frac{1+\cos r\pi}{\pi(r^2-1)}\cos rx$$

$$= \frac{2}{\pi} - \frac{4}{\pi}\left(\frac{\cos 2x}{2^2-1} + \frac{\cos 4x}{4^2-1} + \frac{\cos 6x}{6^2-1} + \ldots\right).$$

9. Express the following function by a cosine and a sine series:

$$f(x) = \begin{bmatrix} x \text{ for } 0 \le x \le \pi/2 \\ \pi - x \text{ for } \pi/2 \le x \le \pi \end{bmatrix}$$

Solution: A function *f(x)* defined over the range $0<x<\pi$ may be represent by a cosine series:

$$f(x) = A_0 + \sum_{r=1}^{\infty} A_r \cos rs$$

where $A_0 = A_0 = \frac{1}{\pi}\int_0^\pi f(x)dx$ and $A_r = \frac{2}{\pi}\int_0^\pi f(x)\cos rx\, dx$

Let us evaluate the constants A_0 and A_r for the given function

$$A_0 = \frac{1}{\pi}\int_0^\pi f(x)dx$$

$$= \frac{1}{\pi}\left[\int_0^{\pi/2} x\, dx + \int_{\pi/2}^{\pi} (\pi - x)\, dx\right]$$

$$= \frac{1}{\pi}\left[\frac{x^2}{2}\right]_0^\pi + \frac{1}{\pi}[\pi x]_{\pi/2}^{\pi} - \frac{1}{\pi}\left[\frac{x^2}{2}\right]_{\pi/2}^{\pi}$$

$$= \frac{\pi}{4.}$$

Again

$$A_r = \frac{2}{\pi}\int_0^\pi f(x)\cos rx\, dx$$

$$= \frac{2}{\pi}\int_0^{2/\pi} x\cos rx\, dx + \frac{2}{\pi}\int_{\pi/2}^{\pi} (x-\pi)\cos rx\, dx$$

$$= \frac{2}{\pi}\left[\frac{1}{r}x\sin rx = \frac{1}{r^2}\cos rx\right]_0^{\pi/2} + 2\left[\frac{1}{r}\sin rx\right]_{\pi/2}^{\pi}$$

$$-\frac{2}{\pi}\left[\frac{1}{r}x\sin rx + \frac{1}{r^2}\cos rx\right]_{\pi/2}^{\pi}$$

$$= \frac{2}{\pi}\left[\frac{1}{r}\frac{\pi}{2}\sin\frac{r\pi}{2} + \frac{1}{r^2}\cos\frac{r\pi}{2} - \frac{1}{r^2}\right]$$

$$+2\left[-\frac{1}{r}\sin\frac{r\pi}{2}\right] - \frac{2}{\pi}\left[\frac{1}{r^2}\cos r\pi - \frac{1}{r}\frac{\pi}{2}\sin\frac{r\pi}{2} - \frac{1}{r^2}\cos\frac{r\pi}{2}\right]$$

$$= -\frac{2}{\pi r^2}\left(1 + \cos r\pi - 2\cos\frac{r\pi}{2}\right)$$

Putting $r = 1, 2, 3,...$

$$A_1 = 0, A^2 = -\frac{8}{2^2\pi}, A_3 = 0, A_4 = 0, A_5 = 0, A_6 = -\frac{8}{6^2\pi}, ...$$

Hence, only the following cosine terms are present:

$$-\frac{8}{\pi}\left(\frac{1}{2^2}\cos 2x + \frac{1}{6^2}\cos 6x + ...\right)$$

$$\therefore f(x) = \frac{\pi}{4} - \frac{8}{\pi}\left(\frac{1}{2^2}\cos 2x + \frac{1}{6^2}\cos 6x + ...\right)$$

Again, a function defined over the range $0<x<\pi$ may also be represented by a sine series:

$$f(x) = \sum_{r=1}^{\infty} B_r \sin rx$$

where $B_r = \frac{2}{\pi}\int_0^{\pi} f(x)\sin rx\, dx.$

Let us evaluate the constant B_r for the given function:

$$B_r = \frac{2}{\pi}\int_0^{\pi} f(x)\sin rx\, dx$$

$$= \frac{2}{\pi}\left[\int_0^{\pi 2} x\sin rx\, dx + \int_{\pi/2}^{\pi} (\pi - x)\sin rx\, dx\right]$$

$$= \frac{2}{\pi}\left[\int_0^{\pi 2} \sin rx\, dx + \int_{\pi/2}^{\pi} \pi\sin rx\, dx - \int_{\pi/2}^{\pi} x\sin rx\, dx\right]$$

$$=\frac{2}{\pi}\left[-\frac{1}{r}x\cos rx+\frac{1}{r^2}\sin rx\right]_0^{\pi/2}+\frac{2}{\pi}\left[\pi\left(\frac{-\cos rx}{r}\right)\right]_{\pi/2}^{\pi}$$

$$-\frac{2}{\pi}\left[-\frac{1}{r^2}x\cos rx+\frac{1}{r^2}\sin rx\right]_{\pi/2}^{\pi}$$

$$=\frac{2}{\pi}\left[-\frac{1}{r}\frac{\pi}{2}\cos\frac{r\pi}{2}+\frac{1}{r^2}\sin\frac{r\pi}{2}\right]+2\left[-\frac{1}{r}r\pi+\frac{1}{r}\cos\frac{r\pi}{2}\right]$$

$$=\frac{2}{\pi}\left[-\frac{1}{r}\pi\cos r\pi+\frac{1}{r}\frac{\pi}{2}\cos\frac{r\pi}{2}-\frac{1}{r^2}\sin\frac{r\pi}{2}\right]$$

$$=\frac{4}{r^2\pi}\sin\frac{r\pi^2}{2}$$

When r is even $\sin r\pi/2=0$, so that terms with odd values of r only are present.

Thus,

$$f(x)=\frac{4}{\pi}\left(\frac{\sin x}{1^2}-\frac{\sin 3x}{3^2}+\frac{\sin 5x}{5^2}-\frac{\sin 7x}{7^2}+\ldots\right).$$

10. Find a series of sines of multiples of x which will represent f(x) in the interval (0, π) where

$$f(x)=\frac{1}{3}\pi\ 0,<x<\frac{1}{3}\pi,$$

$$f(x)=0,\frac{1}{3}\pi<x<\frac{2}{3}\pi,$$

$$f(x)=-\frac{1}{3}\pi,\frac{2}{3}\pi<x<\pi.$$

Can you represent this function by a series of cosines of multiples of x as well ? Explain your answer. Draw graphs of these series and find the sine and cosine series where

$$x=-\frac{1}{3}\pi,-\frac{2}{3}\pi-\pi.$$

Solution: For sine series we have

$$f(x) = \sum_n b \sin nx$$

where $b = \dfrac{2}{\pi}\int_0^{\pi} f(v)\sin nv\, dv$

$$= \frac{2}{\pi}\left[\int_0^{\pi/2\pi} 3\sin nv\, dv + \int_{\pi/3}^{2\pi/3} 0.\sin nv\, dv + \int_{2\pi/3}^{\pi} -\frac{\pi}{3}\sin nv\, dv\right]$$

$$= \frac{2}{3}\left[-\frac{\cos nv}{n}\right]_0^{\pi/3} - \frac{2}{3}\left[-\frac{\cos nv}{n}\right]_{2\pi/3}^{\pi}$$

$$= \frac{2}{3n}\left[1 - \cos\frac{n\pi}{3} + \cos n\pi - \cos\frac{2n\pi}{3}\right]$$

$$= \frac{2}{3n}\left[2\sin^2\frac{n\pi}{6} - 2\sin\frac{5n\pi}{6}\sin\frac{n\pi}{6}\right]$$

$$= \frac{4}{3n}\sin\frac{n\pi}{5}\left[\sin\frac{n\pi}{6} - \sin\frac{5n\pi}{6}\right]$$

$$= -\frac{8}{3n}\sin\frac{n\pi}{6}\sin\frac{n\pi}{3}\cos\frac{n\pi}{2}.$$

$$\therefore\ f(x) = -\frac{8}{3}\sum_{n=1}^{\infty}\frac{1}{3}\sin\frac{n\pi}{6}\sin\frac{n\pi}{3}\cos\frac{n\pi}{2}\sin nx$$

$$= \frac{1}{2}\left[\frac{1}{2}\sin 2x + \frac{1}{2}\sin 4x + \frac{1}{8}\sin 8x + \frac{1}{10}\sin 10x + \ldots\right]$$

For cosine series, we know that $f(x) = a_0 + \Sigma\, a_n \cos nx$
Where

$$a_0 = \frac{1}{\pi}\int_0^{\pi} f(v)\,dv$$

$$= \frac{1}{\pi}\left[\int_0^{\pi/3}\frac{\pi}{3}dv + \int_{\pi/3}^{2\pi/3} 0dv + \int_{2\pi/3}^{\pi} -\frac{\pi}{3}dv = 0\right]$$

$$a_0 = \frac{2}{\pi}\int_0^{\pi} f(v)\cos nv\, dv$$

$$= \frac{2}{\pi}\left[\int_0^{\pi/3}\frac{\pi}{3}\cos nv\, dv + \int_{\pi/3}^{2\pi/3} 0.dv + \int_{2\pi/3}^{\pi} -\frac{\pi}{3}\cos nv\, dv\right]$$

$$\frac{2}{3n}\left[\sin\frac{n\pi}{3}+\sin\frac{2n\pi}{3}\right]=\frac{4}{3n}\sin\frac{n\pi}{2}\cos\frac{n\pi}{6}$$

$$\therefore\ f(x)=\frac{4}{3}\sum_{n=1}^{\infty}\left(\frac{1}{n}\cos\frac{n\pi}{6}\sin\frac{n\pi}{2}\cos nx\right)$$

$$=\frac{2}{\sqrt{3}}\left[\cos x-\frac{1}{5}\cos 5x+7x-\frac{1}{7}\cos 11x+...\right]$$

Graph of the Sine Series: In the interval $(0,\pi)$, the series represent the broken lines given by

$$y\ =\frac{1}{3}\pi\ 0<x<\frac{1}{3}\pi,$$

$$y\ =0\frac{1}{3}\pi<x<\frac{2}{3}\pi,$$

$$y\ =-\frac{1}{3}\pi\frac{2}{3}\pi<x<\pi.$$

Points $x=\frac{1}{3}\pi$ and $x=\frac{2}{3}\pi$ are points of discontinuity.

Graph is as shown in the figure below:

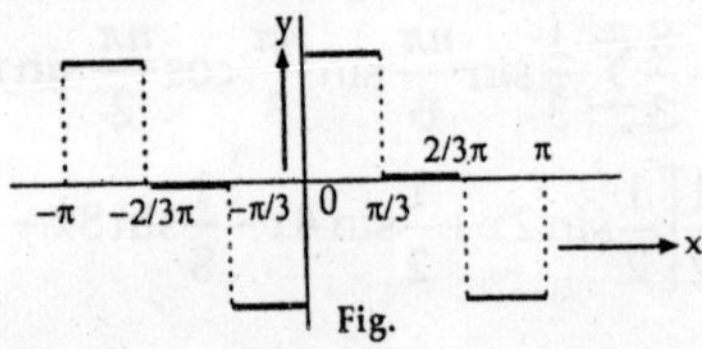

Fig.

There is symmetry in Opposite quadrants; the portion between $(-\pi,\pi)$ repeats indefinitely on both the sides.

For *cosine series.* Graph is represented by same broken lines, but there is symmetry about axis of y.

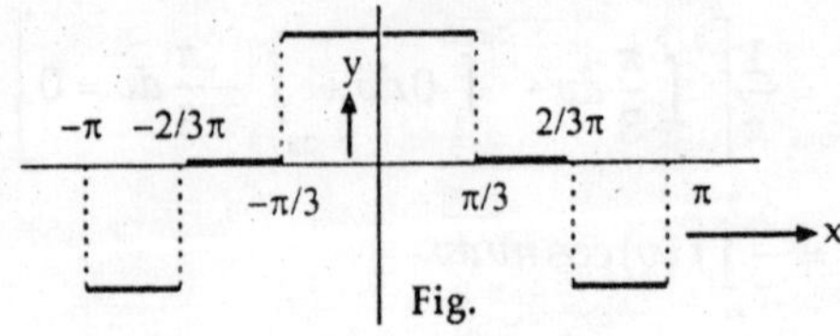

Fig.

11. Find a series of sines and cosined of multiples of x, which will represent f(x) in the interval $-\pi<x<\pi$, ***when***

$$f(x)=\begin{cases} 0, -\pi < x \le 0 \\ \frac{1}{4}\pi x, 0 \le x < \pi \end{cases}$$

and hence deduce that $\frac{1}{2}\pi^2 = 1+\frac{1}{3^2}+\frac{1}{5^2}+\ldots$

Solution: *Let* $f(x) = a_0 + \sum_{n=1}^{\infty} a_n \cos nx + \cos nx + \sum_{n=1}^{\infty} b_n \sin nx$

Again

$$a_0 = \frac{1}{2\pi}\int_{-\pi}^{\pi} f(v)dv = \frac{1}{2\pi}\left[\int_{-\pi}^{\pi} 0.dv + \int_{0}^{\pi}\frac{1}{4}nv\ dv\right]$$

$$= \frac{1}{2\pi}.\frac{\pi}{4}\left[\frac{v^2}{2}\right]_0^{\pi} = \frac{\pi^2}{16},$$

$$a_n = \frac{1}{\pi}\left[\int_{-\pi}^{0} \pi f(v)\cos nv\ dv\right]$$

$$= \frac{1}{\pi}\left[\int_{-\pi}^{0} 0.\cos nv\ dv + \int_{0}^{\pi}\frac{1}{4}nv\cos nv\ dv\right]$$

$$= \frac{1}{4}\left[\frac{v\sin nv}{n}+\frac{\cos nv}{n^2}\right]_0^{\pi} = \frac{\left[(-1)^n - 1\right]}{4n^2}$$

$$b_n = \frac{1}{\pi}\int_{-\pi}^{\pi} f(v)\sin nv\ dv$$

$$= \frac{1}{\pi}\left[\int_{-\pi}^{0} 0.\sin nv\ dv + \int_{0}^{\pi}\frac{1}{4}nv\sin nv\ dv\right]$$

$$= \frac{1}{\pi}\left[\frac{-v\cos nv}{n}+\frac{\sin nv}{n^2}\right]_0^{\pi} = -\frac{(-1)^n\pi}{4\pi}.$$

Therefore

$$f(x) = \frac{\pi^2}{16} + \sum_{n=1}^{x}\frac{\left[(-1)^n - 1\right]}{4n^2}\cos nx - \sum_{n=1}^{\infty}\frac{(-1)^{n\pi}}{4\pi}\sin nx$$

$$= \frac{\pi^2}{16} + \left[-\frac{1}{2}\cos x - \frac{1}{2.3^2}\cos 3x\ldots\right]$$

$$+\left[\frac{\pi}{4}\sin x - \frac{\pi}{4.2}\sin 2x + \frac{\pi}{4.3}\sin 3x\ldots\right]$$

at $x=\pi$ the sum of series is

$$\frac{1}{2}\left[f\left(\pi+0+f\left(\pi-0\right)\right)\right]=\frac{1}{2}\left[0+\frac{1}{4}\pi.\pi\right]=\frac{1}{8}\pi^2$$

$$\therefore \frac{1}{8}\pi^2=\frac{1}{16}\pi^2+\frac{1}{2}\left[1+\frac{1}{3^2}+\frac{1}{5^2}+...\right]$$

(by putting x= π in the above expansion)

or $\frac{1}{8}\pi^2=1+\frac{1}{3^2}+\frac{1}{5^2}+..$

12. *Find a series of cosines of multiples of x, which represents f(x) in the interval 0 and π, where*

$$f(x)=\begin{cases}\frac{1}{4}\pi x, 0\le x<\frac{1}{2}\pi \\ \frac{1}{4}\pi(\pi-x), \frac{1}{2}\pi\le x\le\pi,\end{cases}$$

Solution: For a series of multiples of x, let

$$f(x)=a_0+\sum_{n=1}^{\infty}a_n\cos nx,$$

Again,

$$a_n=\frac{1}{\pi}\int_0^{\pi}f(v)\,dv=\frac{1}{\pi}\left[\int_0^{\pi/2}\frac{1}{4}nv\,dv+\int_{\pi/2}^{\pi}\frac{1}{4}\pi(\pi-v)\,dv\right]$$

$$=\frac{1}{4}\left[\frac{v^2}{2}\right]_0^{\pi/2}+\frac{1}{4}\left[nv-\frac{v^2}{2}\right]_{\pi/2}^{\pi}=\frac{\pi^2}{32}+\frac{\pi^2}{32}=\frac{\pi^2}{16},$$

$$a_n=\frac{2}{\pi}\int_0^{\pi}f(v)\cos nv\,dv$$

$$=\frac{2}{\pi}\left[\int_{\pi}^{\pi/2}\frac{1}{4}\pi v\cos nv\,dv+\int_{\pi/2}^{\pi}\frac{1}{4}\pi(\pi-v)\cos nv\,dv\right]$$

$$=\frac{1}{2}\left[\frac{v\sin nv}{n}+\frac{\cos nv}{n^2}\right]_0^{\pi/2}+\frac{1}{2}\left[\frac{\pi\sin nv}{n}-\frac{v\sin nv}{n^2}-\frac{\cos nv}{n^2}\right]_{\pi/2}^{-\pi}$$

$$=\frac{1}{2}\left[\frac{\pi}{2\pi}\sin\frac{1}{2}n\pi+\frac{\cos\frac{1}{2}n\pi}{n^2}-\frac{1}{n^2}-\frac{\pi\sin\frac{1}{2}n\pi}{n}\right.$$

$$\left.+\frac{\pi}{2n}\sin\frac{1}{2}n\pi-\frac{\cos n\pi}{n^2}+\frac{\cos\frac{1}{2}n\pi}{n^2}\right]$$

$$=\frac{\cos\frac{1}{2}n\pi}{n^2}-\frac{1}{2n^2}-\frac{1}{2n^2}\cos n\pi=\frac{\cos\frac{1}{2}n\pi}{n^2}-\frac{1}{2n^2}\left[1+(-1)^n\right]$$

= 0 when n is odd or a multiple of 4

$=-\frac{2}{n^2}$ when n is even (but not multiple of 4).

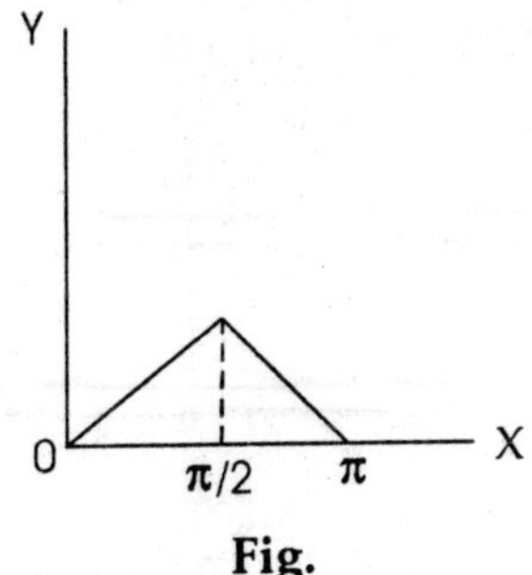

Fig.

$$\therefore\ f(x)=\frac{1}{16}n^2-2\left[\frac{\cos 2x}{2^2}+\frac{\cos 6x}{6^2}+\frac{\cos 10x}{10^2}+\ldots\right]$$

In the interval $(0,\pi)$ the graph of the curve is given by the lines

$$y=\frac{1}{4}\pi x, \qquad 0\le x\frac{1}{2}\pi$$

$$y=\frac{1}{4}(\pi-x), \qquad \frac{1}{2}\pi\le x<\pi.$$

Since cos (–x) = cos x, the curve is symmetrical about the axis of y. The curve from $-\pi$ to π repeats indefinitely in both the directions.

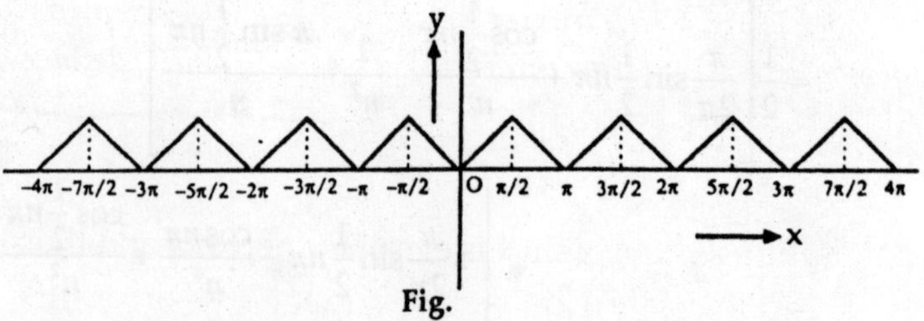

Fig.

13. *Find the Fourier series of the function*

$$f(t)\begin{cases} 0 \text{ when } -2 < t < -1 \\ K \text{ when } -1 < t < 1, T = 4 \\ 0 \text{ when } 1 < t < 2 \end{cases}$$

Solution: Since f is even, we have $b_n = 0$, i.e., we shall have cosine series only,

$$a_0 = \frac{2}{T}\int_0^{T/2} f(t)dt = \frac{1}{2}\int_0^2 f(t)dt$$

Fig.

$$= \frac{1}{2}\left[\int_0^1 f(t)dt + \int_1^2 f(t)dt\right] = \frac{1}{2}\int_0^1 kdt = \frac{k}{2},$$

$$a_n = \int_0^2 f(t)\cos\frac{n\pi}{2}tdt = \int_0^1 k\cos\frac{n\pi}{2}tdt = \frac{2k}{n\pi}\sin\frac{n\pi}{2}$$

Thus, $a_n = 0$ when n is even, an $= \frac{2k}{n\pi}$ when n=1,5,9,... and

$a_n = \frac{-2k}{n\pi}$ when n = 3,7,11,... Hence,

$$f(f) = \frac{k}{\pi + \frac{2k}{\pi}}\left[\cos\frac{1}{2}\cos t - \frac{1}{3}\cos\frac{3}{2}\cos t + \frac{1}{5}\cos\frac{5}{2}\pi t ...\right].$$

Graph of the function is as shown in the figure.

14. ***Obtain Fourier's series for the expansion of f(x) = x sin x in the interval*** $-\pi < x < \pi$. ***Hence, deduce that***

$$\frac{\pi}{4} = \frac{1}{2} + \frac{1}{1.3} - \frac{1}{3.5} + \frac{1}{5.7}$$

Solution: Here x sin x is even function of x and thus in this

$$f(x =) = \frac{1}{\pi}\int_0^{\pi} f(v)\,dv + \frac{2}{\pi}\sum_{n=1}^{\infty}\cos nx \int_0^{\pi} f(v)\cos nv\,dv,$$

$$f(x) = x\sin x; \qquad \therefore f(v) = v\sin v,$$

$$\therefore \sin x = \frac{1}{\pi}\int_0^{\pi} v\sin v\,dv + \frac{2}{\pi}\sum_{n=1}^{\infty}\cos nx \int_0^{\pi} v\sin v\cos nv\,dv.$$

$$\text{Now } \int_0^{\pi} v\sin v\,dv = [-v\cos v + \sin v]_0^{\pi} = \pi$$

$$\text{and } \int_0^{\pi} v\sin v\cos nv\,dv - \frac{1}{2}\int_0^{\pi} v[\sin(n+1)v - \sin(n-1)v]\,dv.$$

$$= \frac{1}{2}\left[-\frac{v\cos(n=1)v}{n+1} + \frac{\sin(n+1)v}{(n+1)^2}\int_0^{\pi} -\frac{1}{2} - \frac{v\cos(n-1)v}{n-1} + \frac{\sin(n-1)v}{(n-1)^2}\right]_0^{\pi}$$

$$= \frac{\pi}{2}\left[-\frac{\cos(n-1)\pi}{n-1} - \frac{\cos(n+1)\pi}{n+1}\right]$$

$$= \frac{\pi}{2}\left[-\frac{\cos n\pi}{n-1} - \frac{\cos n\pi}{n+1}\right]$$

$$= \frac{\pi\cos nx}{1-n^2}\ when\ n > 1.$$

When n = 1, we have

$$\int_0^{\pi} v\sin v\cos v\,dv = \frac{1}{2}\int_0^{\pi} v\sin 2v\,dv$$

$$= \frac{1}{2}\left[-\frac{v\cos 2v}{2} + \frac{\sin 2v}{4}\right]_0^{\pi} = -\frac{\pi}{4}$$

$$\text{Therefore } x\sin x = 1 + \frac{2}{\pi}\left(-\frac{\pi}{4}\cos x + \sum_{n=2}^{\infty}\frac{\pi\cos n\pi}{1-n^2}\cos nx\right)$$

$$=1+2\left[-\frac{1}{4}\cos x-\frac{1}{1.3}\cos 2x+\frac{1}{2.4}\cos 3x+\frac{1}{3.5}\cos 4x+\ldots\right].$$

Putting $x=\frac{1}{2}\pi$, we get

$$\frac{\pi}{2}=1+2\left\{\frac{1}{1.3}-\frac{1}{3.5}+\frac{1}{5.7}-\ldots\right\}$$

or $\frac{\pi}{4}=\frac{1}{2}+\frac{1}{1.3}=\frac{1}{3.5}+\frac{1}{5.7}-\ldots$

15. *Find a series of sines and cosines of multiples of x, which will represent $x+x^2$ in the interval $-\pi<x<\pi$. Further deduce that*

$$\frac{\pi^2}{6}=1+\frac{1}{2^2}+\frac{1}{3^2}+\ldots$$

Solution. $f(x)=x=x^2$; $\therefore f(v)=v+v^2$

We have

$$f(x)=\frac{1}{2\pi}\int_{-\pi}^{\pi}\left(v+v^2\right)dv+\frac{1}{\pi}\sum_{n=1}^{\infty}\cos nx\int_{-\pi}^{\pi}\left(v+v^2\right)\cos nv\,dv$$

$$+\frac{1}{\pi}\sum_{n=1}^{\infty}\sin nx\int_{-\pi}^{\pi}\left(v+v^2\right)\sin nv\,dv.$$

Now

$$a_0=\frac{1}{2\pi}\int_{-\pi}^{\pi}\left(v+v^2\right)dv=\frac{1}{2\pi}\left[\frac{v^2}{2}+\frac{v^3}{3}\right]_{-\pi}^{\pi}=\frac{\pi^2}{3},$$

$$a_n=\frac{1}{\pi}\int_{-\pi}^{\pi}\left(v+v^2\right)\cos nv\,dv=\frac{2}{\pi}\int_0^{\pi}v^2\cos nv\,dv.$$

(other interal vanishes)

$$=\frac{2}{\pi}\left[\left(\frac{v_2\sin nv}{n}\right)_0^{\pi}-\frac{2}{n}\int_0^{\pi}v\sin nv\,dv\right]$$

$$=-\frac{4}{n\pi}\left[-\frac{v\cos nv}{n}+\frac{\sin nv}{n^2}\right]_0^{\pi}=\frac{4}{n^2}\cos n\pi,$$

$$b_n = \frac{1}{\pi}\int_{-\pi}^{\pi}(v+v^2)\sin nv\,dv = \frac{2}{\pi}\int_0^{\pi} v\sin nv\,dv.$$

$$= \frac{2}{\pi}\left[\left(\frac{v\cos nv}{n}+\frac{\sin nv}{n^2}\right)\right]_0^{\pi} = -\frac{2}{n}\cos n\pi.$$

$$\therefore x+x^2+\frac{\pi 2}{3}+4\sum_{n=1}^{\infty}\frac{\cos n\pi}{n^2}\cos nx - 2\sum_{n=1}^{\infty}\frac{\cos n\pi}{n}\sin nx$$

$$= \frac{\pi^2}{3}-4\left\{\cos x-\frac{1}{2^2}\cos 2x+\frac{1}{3^2}\cos 3x\ldots\right\}$$

$$+2\left\{\sin x-\frac{1}{2}\sin 2x+\frac{1}{3}\sin 3x\ldots\right\}$$

Second Part

At $x = \pm\pi$, the sum of series

$$= \frac{1}{2}\left[f(-\pi+0)+\pi-0\right]$$

$$= \frac{1}{2}\left[-\pi+\pi^2+\pi+\pi^2\right] = \pi^2.$$

Putting $x = \pi$ in (A), we have

$$f(\pi) = \pi^2 = \frac{\pi^2}{3}+4\left\{1+\frac{1}{2^2}+\frac{1}{3^2}+\ldots\right\}$$

or $$\frac{\pi^2}{6} = 1+\frac{1}{2^2}+\frac{1}{3^2}+\ldots$$

EXERCISE

1. Find the Fourier series of the following function f(x) having a period 2π

$$f(x) = \begin{cases} x \text{ for } -\pi/2 < x < \pi/2 \\ 0 \text{ for } \pi/2 < x < 3\pi/2 \end{cases}.$$

2. Expand the following function in a Fourier series:

$$f(x) = \begin{cases} 1 \text{ for } -\pi < x < 0 \\ 2 \text{ for } 0 < x < \pi \end{cases}.$$

3. Expand by Fourier's series, the following function:

$$f(x)=\begin{cases} 1 \text{ for } -\pi/2<x<\pi/2 \\ -1 \text{ for } \pi/2<x<3\pi/2 \end{cases}.$$

4. Expand the following functions in a cosine series and a sine series:

   (i) $f(x)=x$ when $0<x<\pi$.

   (ii) $f(x)=\pi-x$ when $0<x<\pi$.

ANSWERS

1. $f(x)=\frac{2}{\pi}\sin x+\frac{1}{2}\sin 2x-\frac{2}{9\pi}\sin 3x-\frac{1}{4}\sin 4x+\frac{2}{25\pi}\sin 5x.$

2. $f(x)=\frac{3}{2}+\frac{2}{\pi}\left(\sin x+\frac{\sin 3x}{3}+\frac{\sin 5x}{5}+\ldots\right)$

3. $f(x)=\frac{4}{\pi}+\left(\cos x-\frac{1}{3}\cos 3x+\frac{1}{5}\cos 5x-\ldots\right)$

4. (i) $\frac{\pi}{2}-\frac{4}{\pi}+\left(\cos x-\frac{\cos 3x}{3^2}+\frac{\cos 5x}{5^2}+\ldots\right),$

   $2\left(\sin x-\frac{\sin 2x}{2}+\frac{\sin 3x}{3}-\ldots\right)$

   (ii) $2\left(\sin x-\frac{\sin 2x}{2}+\frac{\sin 3x}{3}+\ldots\right),$

   $\frac{\pi}{2}-\frac{4}{\pi}+\left(\cos x+\frac{\cos 3x}{3^2}+\frac{\cos 5x}{5^2}+\ldots\right).$

Significance of Temperature

An Ideal Gas: An ideal gas is on which strictly obeys the Boyle's law and Charle's law under all conditions of temperature and pressure.

Its molecules do not exert force of attraction on one another. The internal energy is, therefore, wholly kinetic, depending on the temperature only.

Deviation of Real Gases from Gas Laws—Boyle Temperature: No real gas strictly follows the gas laws. Gases which are easily liquefied, such as carbon-dioxide, ammonia, etc. show marked deviation from gas laws.

The so-called permanent gases, such as oxygen, nitrogen, hydrogen, etc. show little deviation at ordinary temperatures and pressures, but marked deviation at low temperatures and high pressures.

Experiments to investigate the actual behaviour of real gases and the amount of their deviation from gas laws have been extensively performed.

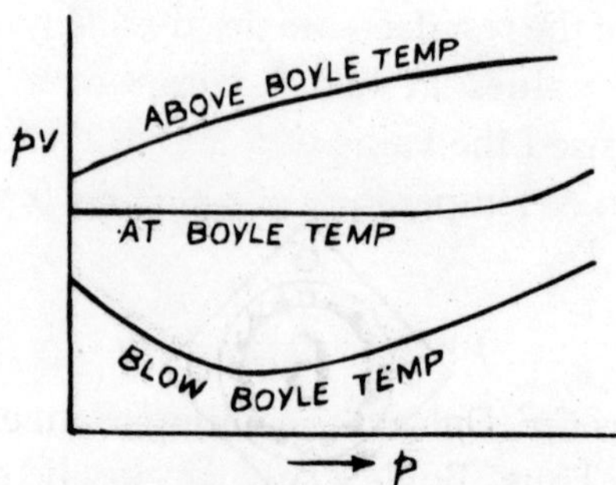

The results of these experiments are expressed by plotting the product pV against p for different temperatures. The general nature of curves is the same for all gases. We draw following conclusions from these curves:

(i) For each gas there is a particular temperature at which the product pV remains constant as the pressure is increased. This is called the "Boyle temperature". At very high pressure, pV begins to increase.

(ii) Below the Boyle temperature, pV first decreases to a minimum as p is increased, and then begins to increase.

(iii) Above the Boyle temperature, pV increases with increase in p from the very beginning.

The Boyle temperatures for He, H_2 and N_2 are – 250°C, – 164°C and 50°C, respectively. Therefore, at ordinary temperatures, for He and H_2 the product pV increases with p; while for N_2 it first decrease to a minimum and then increases.

In order to represent analytically the behaviour of real gases, Kamerlingh Onnes introduced the following empirical relation:

$$pV = A + Bp + Cp^2 + Dp^3 + ...,$$

where the constants A, B, C, D ... are characteristics of a given gas at a given temperature and are called "virial coefficients". For 1 mole of the gas, the first coefficient A is equal to RT and the magnitudes of the rest decrease progressively. The constant B has negative values at low temperatures, and as the temperature is raised the value of B passes through zero and becomes positive. At temperature at which $B = 0$, we have very nearly,

$$pV = A \text{ (constant)},$$

because the terms Cp^2, Dp^3, etc. are unimportant except at very high pressures. Thus, Boyle's law is closely obeyed upto sufficiently high pressures. This is the "Boyle temperature". Below this temperature, B is negative so that pV decreases with increase in p, but with further increase in p the term Cp^2 becomes effective and ultimately predominates so that pV begins to increase. Above the Boyle temperature, B is positive so that pV increases with increase in p from the very beginning.

Andrew's Experiment

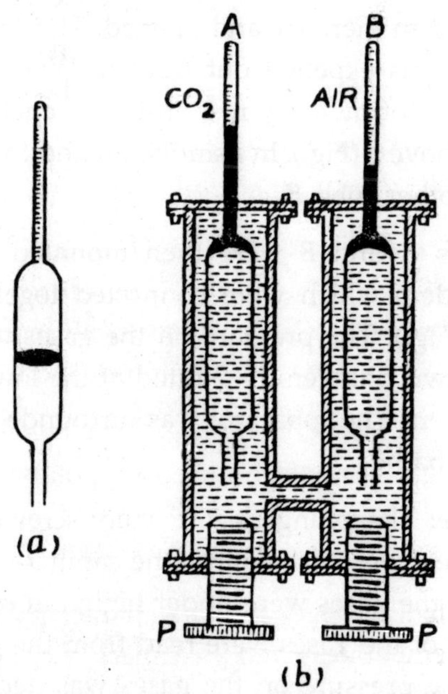

Ordinarily, liquids and gases are two different states of matter. The change from one state to the other takes place *abruptly* with a sharp distinction between the two states. An experiment performed by Cagniard de la Tour in 1822, however, indicated that under certain circumstances the liquid and gaseous states appear to merge into one, and the change from one state to the other takes place *continuously*. This continuity of state was well established by Andrews' experiments on carbon dioxide in 1863. Andrews' experiments also made clear that why in the earlier attempts the gases CO_2, ammonia, etc. liquefied under pressure but N_2, O_2, H_2, etc. did not liquefy even under very high pressures.

Apparatus: The scheme of Andrews' apparatus is shown in Fig. A thick-walled glass tube, the upper part of which was a graduated capillary, was selected. Dried carbon-dioxide was passed through the tube for several hours to remove all traces of air. The ends of the tube were then sealed. The lower end was immersed in mercury and opened. The tube was heated so that some gas expelled out from it. When the tube was cooled, a pellet of mercury rose in it and enclosed some gas in the space above it (Fig.). In a similar manner, air was enclosed in a similar other tube *B*.

The tubes *A* and *B* were then mounted in two strong copper cylinders which were connected together and filled with water (Fig.). The pressure on the gases was applied by screwing in two plungers *P*, *P* fitted at the lower ends of the copper cylinders. The apparatus was surrounded by a constant-temperature bath.

Procedure: The plungers *P*, *P* were screwed in until the mercury pellets were driven into the capillary portions of the tubes so that the gases were under high, but *equal*, pressures. The volumes of the gases were read from the graduations on the tubes. The pressure on the gases was deduced from the volume of air in tube *B*, assuming Boyle's law to apply for air. By increasing the pressure in suitable steps, a series of corresponding values of the volume and pressure of CO_2 were obtained. The experiment was repeated at different temperatures.

The results were represented by means of curves drawn by plotting the pressure of CO_2 against the corresponding volume at different temperatures. These curves are called "isothermals".

Discussion of Results: Let us first consider the isothermal at temperature 13.1°C, starting with carbon dioxide in the state represented by the point *A*. As the pressure is increased, the volume is first decreased roughly according to Boyle's law, as shown by the part *AB* of the isothermal. At *B*, which corresponds to about 50 atm pressure, there is a discontinuity in the curve. At this point the gas has begun to liquefy. Now, as the liquefaction continues, the volume decreases but the pressure remains constant as indicated by the horizontal part *BC*. At C the carbon dioxide is completely liquefied. As the pressure is further increased, the volume remains practically undiminished (part *CD)*, thus confirming that the carbon dioxide is wholly liquid, since liquids are almost incompressible.

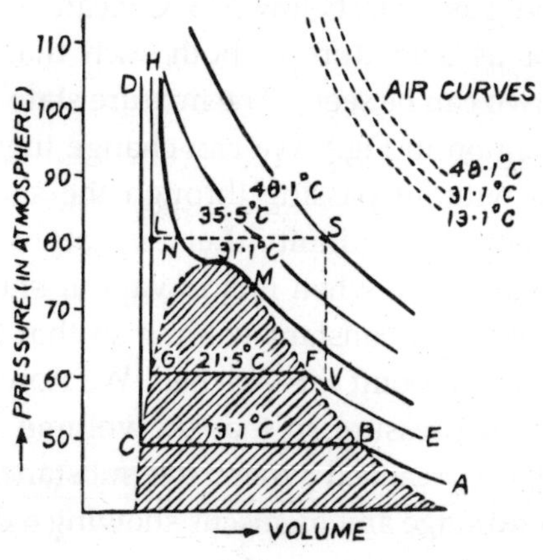

The isothermal for 21.5°C is, in general, similar to that for 13.1°C, but the liquefaction now needs a higher pressure (about 60 atm) corresponding to the point *F*. Also its horizontal part *FG*, over which the vapour and liquid states coexist, is shorter than *BC*. This part continuously decreases, as the temperature is raised, and disappears in the isothermal for 31.1°C. This isothermal has no discontinuities (such as *B* and *F*) but has a continuous kink *MN* over which the volume decreases very rapidly with increase in pressure, but there is no evidence of liquefaction. Even this kink becomes less marked at 35.5°C and disappears at 48.1°C when the isothermal becomes similar to those for air shown at the top of the diagram.

It is thus clear that however great the applied pressure, liquefaction of carbon-dioxide is not possible at temperatures above 31.1°C. Repeated experiments showed that the highest temperature at which it is just possible to liquefy carbon-dioxide is 30.9°C. This is called the "critical temperature", T_c. The pressure required to produce liquefaction at this temperature is called the "critical pressure" p_c and the volume occupied by one gram of the gas under these pressure and temperature is called the "critical volume" V_c.

Andrews' curves illustrate another important fact regarding liquid and gaseous states of a substance. Above 30.9°C, carbon-dioxide exists only as a gas; below 30.9°C it can exist as a liquid or a vapour, or as a mixture of both such that a meniscus separating the two can be seen. (The mixture state corresponds to the shaded region in Fig.). We can change the vapour into the liquid state without passing through the shaded region, thus avoiding any discontinuity. Let us start with the state *V* (Fig.) of the substance when it is in vapour state at 21.5°C. We heat the vapour at constant volume so that the pressure is increased until the point *S* is reached. We now cool it upto 21.5°C at constant pressure so that the volume is decreased until the point *L* is reached where the substance is entirely liquid. Thus, at no stage any meniscus showing a discontinuity appears. Such a continuous transition between the two states establishes the *continuity of state.* In other words, the two states are only distant stages of a long series of continuous physical changes.

Equation of State of an Actual Gas: The ideal gas equation $pV = RT$ does not correctly represent the experimental behaviour of real gases. It is based on two assumptions, namely:

(i) The molecules of a gas are infinitesimally small in size.

(ii) They do not exert attraction on one another.

Actually the molecules of a real gas do possess a finite size and also attract one another. Hence, the equation of state of an actual gas will be different from that of an ideal gas.

Derivation of Van der Waals' Equation of State: Van der Waals modified the ideal gas equation by correcting it for the finite size of the molecules and for their mutual attraction.

The equation of state for 1 mole of an ideal gas occupying a volume V_i at a pressure p_i is,

$$P_i V_i = RT. \qquad \text{...(i)}$$

Let us now consider 1 mole of an *actual* gas occupying an observed volume V at an observed pressure p, and apply corrections.

Correction for Finite Size of Molecules: The molecules of a gas occupy a finite volume in space. Therefore, the volume actually available for the molecules to move freely is *less* than the volume of the gas. Thus, a real gas occupying a volume V is equivalent to an ideal gas (whose molecules are of negligible size) of *smaller* volume V_i, such that,

$$V_i = V - b. \qquad \text{...(ii)}$$

b is a constant for 1 mole of the gas and is called the 'co-volume'. It has been shown that b is four times the actual volume of the molecules present in the gas, because each molecule is surrounded by a sphere of space within which no other molecule can enter.

Correction for Intermolecular Attraction: A molecule in the body of the gas is uniformly surrounded by other molecules. Hence, it is attracted equally in all directions, *i.e.* it experiences no resultant force. A molecule near the wall of the containing vessel, however, experiences an *inward* pull due to unbalanced attractions of the molecules behind it, as shown in Fig. On account of this pull, the molecule collides against the wall with a *smaller* momentum, *i.e.* the effective pressure of the gas is reduced. Thus, the observed pressure p exerted by a real gas is less than the pressure P_i which would have been exerted by an ideal gas (in which there is no molecular attraction). Therefore

$$p = p_i - \beta$$

or $$p_i = p + \beta. \quad \text{...(iii)}$$

The reduction in pressure, β, depends upon the inward pull on a molecule near the wall due to the attraction of molecules behind it, as well as on the number of molecules colliding against unit area of the wall per second. Each of these factors is proportional to the density of the gas (number of molecules in unit volume of the gas). Therefore, β will be directly proportional to the square of the density of the gas or inversely to the volume occupied by a given mass of the gas. Thus,

$$\beta = \frac{a}{V^2},$$

where a is a constant for 1 mole of the gas a/V^2 is called 'internal pressure'. Substituting this value of β is eq. (iii), we get,

$$p_i = p + \frac{a}{V^2}, \quad \text{...(iv)}$$

Substituting for V_i and p_i from eq. (ii) and (iv) in eq. (i), we obtain,

$$\boxed{\left(p+\frac{a}{V^2}\right)(V-b)=RT.} \quad ...(v)$$

This is Van der Waals' equation of state for 1 mole of an actual gas.

For μ moles of gas, the eq. would be,

$$\left(p+\frac{a\mu^2}{V^2}\right)(V-\mu b)=\mu RT.$$

Comparison of Theoretical Curves with Experimental Curves: In order to compare the Van der Waals' equation with experimental results, let us put the equation in the form:

$$p = \frac{RT}{V-b}-\frac{a}{V^2}$$

Let us now substitute the values of 'a' and 'b' for a gas in the equation and plot p against V at different constant temperatures. Fig. shows such theoretical isothermals drawn for carbon-dioxide at the same temperatures at which Andrews had drawn his isothermals from experimental data. Let us now compare Van der Waals' theoretical isothermals (Fig.) with Andrews' experimental isothermals (Fig.).

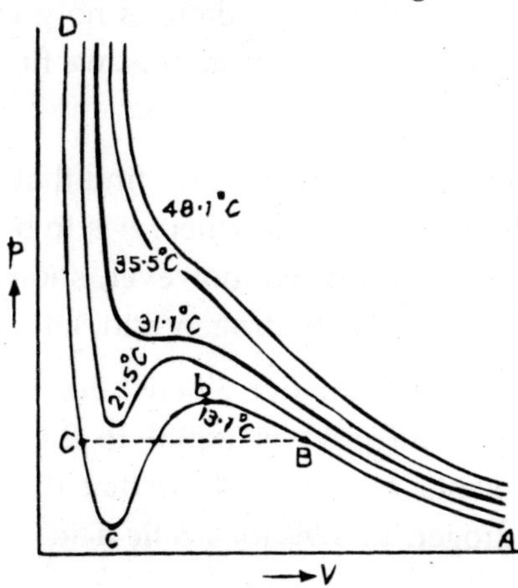

It is observed that the theoretical isothermals have a general resemblance with the experimental isothermals, specially at higher temperatures. At lower temperatures, however, we see remarkable disagreement. Let us consider the isothermal for 13.1°C.

The theory gives a continuous curve *BbcC* (Fig.) having a maximum at *b* and a minimum at *c*, whereas the experiment gives a straight line *BC* (Fig.). The disagreement can, however, be explained.

The parts *Bb* and *Cc* of the theoretical isothermal represent a *super-cooled* vapour and a *super-heated* liquid, respectively. Ordinarily, these states of vapour and liquid are not obtained (hence Andrews did not obtain them), but they can be obtained under very special conditions by using very pure vapour and liquid.

The part *bc* in the theoretical isothermal shows decrease in volume with decrease in pressure, which is essentially an unstable condition never realisable in practice. Hence, this part is not obtained in the Andrews' experimental isothermal.

Defects of Van der Waals' Equation: The Van der Waals' equation has fairly good qualitative agreement with the experimental observations, but there is no good quantitative agreement everywhere. Apart from this, we find the following drawbacks:

(i) According to the equation we find that for any gas the critical volume $V_c = 3b$, where b is to be taken for that very gas. Experiments, however, show that V_c varies between 2.8 b for hydrogen and 1.41 b for argon.

(ii) According to Van der Waals' equation the critical coefficient RT_c/p_cV_c is equal to 2.67 for all gases. Experiments, however, show that it varies from 3.03 for hydrogen to 4.99 for acetic acid.

Hence, we conclude that even the Van der Waals' equation describes the behaviour of real gases *only approximately*.

Variation of a and b with Temperature: Accurate experiments show that Van der Waals' constants a and b are not strictly constants, but vary with temperature.

The variation of a is due to the action of the attractive forces in changing the concentration of the molecules. Obviously, the change in molecular concentration will be less effective at higher kinetic energy of the molecules, *i.e.* at higher temperature of the gas. Thus, a is a function of T and decreases with rise in T.

The variation of b with temperature can also be explained. Van der Waals assumed a constant diameter σ for the molecules. The molecular structure, however, shows that molecules have some softness with a slightly extended force field. Therefore, the smallest distance of approach between the centres of two molecules varies with the energy of the molecules, *i.e.* with the temperature of the gas. At higher temperatures the collisions will be more violent and the molecules will approach each other more closer. Hence, the effective diameter of the molecules, or b, will be a slowly-varying function of temperature.

As the Volume per Mole of a Gas increases, the Van der Waals' Equation Tends to the Equation of State of an Ideal Gas: The Van der Waals' equation for 1 mole of a gas is,

$$\left(p+\frac{a}{V^2}\right)(V-b)=RT.$$

If we increase the volume per mole (V) of the gas (*i.e.* decrease the pressure), the internal pressure a/V^2 decreases rapidly and also the co-volume b becomes negligibly small. Thus, the equation tends to the equation of state of an ideal gas ($pV = RT$).

Limitation of Ideal Gas Equation: The ideal gas equation applies to real gases at low pressures and high temperatures only. In general, real gases sufficiently deviate from it. This deviation has been described in Q. 1, by means of the curves between pV and p drawn in Fig.

Explanation by Van der Waals' Equation: According to the ideal gas equation $pV = RT$, the product pV should remain constant at constant temperature. Hence, the curves between pV and p should be parallel to the pressure-axis. Actually, the curves are not so. Hence, the equation $pV = RT$ does not correctly represent the behaviour of a real gas. The Van der Waals' equation, however, explains it. The equation is,

$$\left(p+\frac{a}{V^2}\right)(V-b) = RT$$

or $$pV - pb + \frac{a}{V} - \frac{ab}{v^2} = RT$$

or $$pV = RT + pb - \frac{a}{V} + \frac{ab}{V^2}$$

Since the terms a/V and ab/V^2 are very small, we can make the approximate substitution of p/RT for $1/V$. Then, we have,

$$pV = RT + pb - \frac{ap}{RT} + \frac{abp^2}{R^2T^2}$$

or $$pV = RT + p\left(b - \frac{a}{RT}\right) + \frac{abp^2}{R^2T^2}.$$

At low pressures, the term $abp^2/R^2\,T^2$ is insignificant since a and b are small. Hence,

$$pV = RT + p\left(b - \frac{a}{RT}\right).$$

At low temperature, $\left(b - \frac{a}{RT}\right)$ will be negative. Therefore, as p increases, the product pV decreases. But as p increases, the term abp^2/R^2T^2 becomes more significant and beyond a particular value of p, the product pV begins to increase with p. This explains the lowest curve of Fig.

At high temperature, a/RT becomes small so that the term $\left(b - \frac{a}{RT}\right)$ is positive. Hence, pV increases with increase in p from the very beginning. This explains the upper curve of Fig.

When the temperature is such that,

$$b - \frac{a}{RT} = 0 \text{ or } T = \frac{a}{bR},$$

pV remains constant for a sufficient increase in pressure until the term $\frac{abp^2}{R^2T^2}$ becomes significant. This is the "Boyle temperature". This explains the middle curve of Fig.

The Van der Waals' equation of state,

$$\left(p + \frac{a}{V^2}\right)(V - b) = RT$$

has been derived in Q. 3 (b).

Critical Constants in Terms of 'a' and 'b': The three critical constants of a gas are: the critical temperature T_c, the critical pressure p_c and the critical volume V_c. To determine their values, let us write the Van der Waals' equation in the following form:

$$p = \frac{RT}{V - b} - \frac{a}{V^2}. \qquad \text{...(i)}$$

If p is plotted against V at different constant values of T, a set of isothermals is obtained (Fig.). The isothermal through the critical point X is horizontal at X, *i.e.* the slope of the isothermal at X is zero. Therefore, at X,

$$\left(\frac{dp}{dV}\right)_T = 0.$$

Also, the curvatures of this isothermal on the two sides of the critical point X are opposite, *i.e.* the point X is a point of inflexion. Therefore, at X,

$$\left(\frac{d^2p}{dV^2}\right)_T = 0.$$

Applying these two conditions to the Van der Waals' equation (i), we get,

$$\left(\frac{dp}{dV}\right)_T = -\frac{RT}{(V-b)^2}+\frac{2a}{V^3}=0 \quad \text{...(ii)}$$

and

$$\left(\frac{d^2p}{dV^2}\right)_T = \frac{2RT}{(V-b)^3}-\frac{6a}{V^4}=0 \quad \text{...(iii)}$$

But at the critical point; $p = p_c$, $V = V_c$ and $T = T_c$. Hence, eq. (i), (ii) and (iii) give,

$$p_c = \frac{RT_c}{(V_c-b)}-\frac{a}{V_c^2}, \quad \text{...(iv)}$$

$$\frac{RT_c}{(V_c-b)^2} = \frac{2a}{V_c^3} \quad \text{...(v)}$$

and

$$\frac{2RT_c}{(V_c-b)^3} = \frac{6a}{V_c^4}. \quad \text{...(vi)}$$

Dividing eq. (v) by eq. (vi), we get,

$$\frac{1}{2}(V_c-b) = \frac{1}{3}V_c$$

or

$$\boxed{V_c = 3b.} \quad \text{...(vii)}$$

Substituting $V_c = 3b$ in eq. (v), we get,

$$\frac{RT_c}{4b^2} = \frac{2a}{27b^3}$$

or

$$\boxed{T_c = \frac{8a}{27bR}.} \quad \text{...(viii)}$$

Substituting $V_c = 3b$ and $T_c = \dfrac{8a}{27bR}$ in eq. (iv), we get,

$$p_c = \frac{R}{2b}\left(\frac{8a}{27bR}\right) - \frac{a}{9b^2}$$

or $$\boxed{p_c = \frac{a}{27b^2}.} \qquad \ldots\text{(ix)}$$

$$\therefore \qquad \frac{RT_c}{p_c V_c} = \frac{R8a}{27bR}\left(\frac{27b^2}{a}\right)\frac{1}{3b} = \frac{8}{3} = 2.67.$$

$\frac{RT_c}{p_c V_c}$ is called the "critical coefficient" and is the same for all gases.

Determination of *'a'* and *'b'*: From eq. (viii) and (ix), we have,

$$\frac{T_c^2}{p_c} = \left(\frac{8a}{27bR}\right)^2 \frac{27b^2}{a}$$

or $$a = \frac{27R^2}{64}\left(\frac{T_c^2}{p_c}\right).$$

Again from eq. (viii) and (ix), we have,

$$\frac{T_c}{p_c} = \frac{8a}{27bR} \times \frac{27b^2}{a}$$

or $$b = \frac{R}{8}\left(\frac{T_c}{p_c}\right).$$

Thus, *'a'* and *'b'* can be determined if T_c and p_c are known. T_c and p_c can be obtained directly from experiments.

Critical Constants: The three critical constants of a gas are critical temperature, critical pressure and critical volume.

Critical Temperature: The critical temperature of a gas is the highest temperature at which the gas can be liquefied by pressure, but above which no liquefaction is possible, however great the pressure.

Critical Pressure: The pressure required to liquefy a gas at- its critical temperature is called the critical pressure.

Critical Volume: The volume occupied by 1 gram of the gas at the critical temperature and the critical pressure is called the critical volume.

Determination of Critical Constants: The critical temperature and the critical pressure can be measured by a simple apparatus used by Cagniard de la Tour. It consists of a strong bent tube closed at both ends (Fig.).

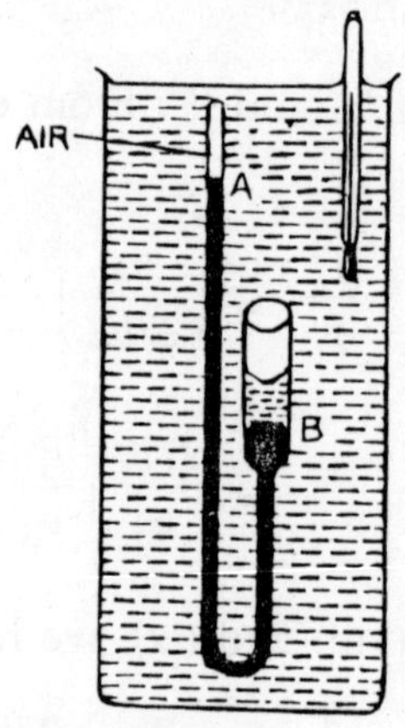

It is filled with mercury from *A* to *B*. The space above *A* contains air while the space above *B* contains the experimental liquid and its vapour. The tube is placed in a thermostat which can be maintained at any desired temperature. At ordinary temperatures the liquid meniscus in *B* is sharp and clear. As the temperature is raised, the meniscus becomes flatter and less distinct and finally disappears altogether. The temperature at which this happens is noted. Now the temperature is lowered and when the meniscus just reappears, the temperature is again noted. The mean of these two temperatures gives the critical temperature. The vapour pressure at critical temperature is estimated by the volume of air above *A* and applying Boyle's law. This is the critical pressure.

The critical volume is measured by a method based on the *law of rectilinear diameters.* In this method the densities of the liquid and its saturated vapour are measured at various temperatures upto as near the critical temperature as possible. On plotting density against temperature, a curve is obtained (Fig.).

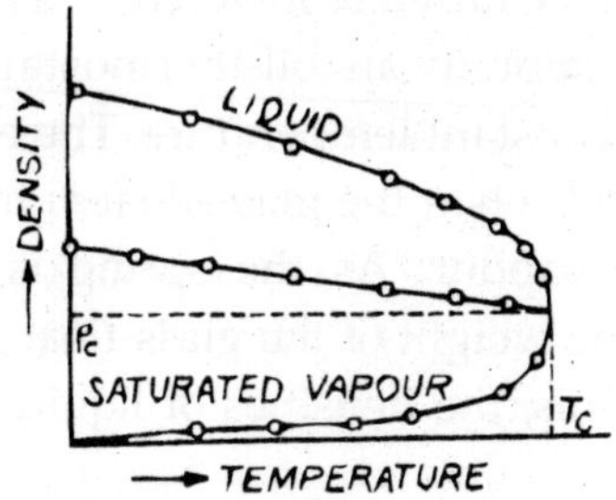

It shows that the densities in the liquid and vapour states come closer to each other and become equal at the critical temperature where the two halves of the curve meet. The mean of the liquid and vapour densities at various temperatures is found out and plotted against the corresponding temperature. The resulting curve is almost a straight line called the 'rectilinear diameter' passing through the critical temperature. The point where the rectilinear diameter meets the curve corresponds to the critical density. From this the critical volume is calculated.

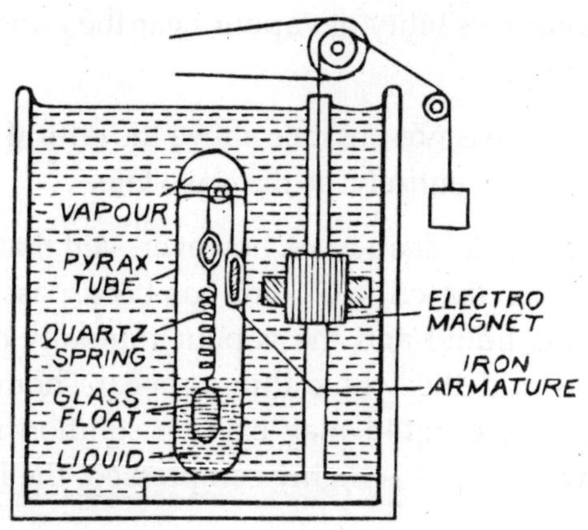

The densities can be measured by a method suggested by Maas. The liquid with its vapour is contained in a small heavy-walled pyrex glass tube (Fig.).

A calibrated quartz spring with a glass-float of known mass and volume is suspended in the tube by means of a pulley. The float can be raised or lowered by a magnetic device. The apparatus is kept in an oil-thermostat which can be maintained at any constant temperature. The extensions of the spring are measured when the glass-float is in the liquid and when it is in the vapour. As the spring is calibrated, the buoyancy-loss in the weight of the glass-float in the two cases is obtained. From this, the densities of liquid and vapour are calculated.

Properties of Fluids near Critical Point: The experiments of Cagniard de la Tour, Andrews and others have revealed many interesting properties of fluids near the critical point.

(i) There is no distinction between the liquid and vapour states, *i.e.* the whole content seems homogeneous.

(ii) The densities of the liquid and the vapour come closer to each other and become equal at the critical point.

(iii) The surface tension vanishes.

(iv) The compressibility of vapour near the critical point is very high.

(v) Liquid state is not possible above the critical point. The substance is entirely in the gas state.

Cailletet and Colardeau have, however, said that the liquid state persists after the critical point has been passed. At the critical point the liquid and the vapour mutually dissolve in all proportions and the whole thing appears homogeneous. This means that the liquid may exist in solution in its own vapour, and when a gas is highly compressed, the liquid may be present.

Calendar's experiment on water has shown that the meniscus vanishes at 374°C, but the properties of liquid and vapour do not become identical before 380°C. This indicated the existence of a critical region rather than a critical point.

Boyle Temperature: According to the ideal gas equation $pV = RT$, the product pV for a given gas must remain constant at constant temperature. Real gases deviate from this. For them, pV varies as p increases (Fig.). This variation is different at different temperatures.

However, at a particular temperature, pV remains constant over a sufficient range of p, provided p is not too large. This temperature is known as "Boyle temperature" and is different for different gases.

Let us compute Boyle temperature for a Van der Waals' gas:

$$\left(p+\frac{a}{V^2}\right)(V-b) = RT$$

or

$$pV = RT+pb-\frac{a}{V}+\frac{ab}{V^2}.$$

Since the terms a/V and ab/V^2 are very small, we can make the approximate substitution of p/RT for $1/V$. Then, we have

$$pV = RT+pb-\frac{ap}{RT}+\frac{abp^2}{R^2T^2}$$

$$pV = RT+p\left(b-\frac{a}{RT}\right)+\frac{abp^2}{R^2T^2}.$$

At low pressures, the term $abp/R^2 T^2$ is insignificant since a and b are small. Hence,

$$pV = RT+p\left(b-\frac{a}{RT}\right).$$

Differentiating pV with respect to p, we get,

$$\frac{d(pV)}{dp} = b-\frac{a}{RT}.$$

Now, at Boyle temperature T_B (say), the product pV remains constant over a sufficient range of p, that is, at $T = T_B$,

$$\frac{d(pV)}{dp} = 0$$

Substituting this in the last expression, we get,

$$b - \frac{a}{RT_B} = 0$$

or

$$\boxed{T_B = \frac{a}{bR}.}$$

We know that the critical temperature of a Van der Waals' gas is given by,

$$T_c = \frac{8a}{27bR}.$$

The above two relations give,

$$\boxed{T_B = \frac{27}{8} T_c.}$$

This is the relation between Boyle temperature and critical temperature.

Shortcoming of Van der Waals' Equation: The Van der Waals' equation of state,

$$\left(p + \frac{a}{V^2}\right)(V - b) = RT,$$

is an accurate representation of the behaviour of real gases. But the constants a and b depend on the properties of the molecules of the particular gas and so are different for different gases. This shortcoming has been removed in the 'reduced equation of state'.

Reduced Equation of State: If p, V, T are actual pressure, volume and temperature of a gas; and p_c, V_c, T_c the critical pressure, volume, and temperature then p/p_c, V/V_c, T/T_c are defined as the reduced pressure p_r, reduced volume V_r and reduced temperature T_r, respectively. Thus, we have,

$$p = p_r p_c,$$
$$V = V_r V_c,$$
and
$$T = T_r T_c.$$

Putting these values of p, V and T in Van der Waals' equation, we get,

$$\left(p_r p_c + \frac{a}{V_r^2 V_c^2}\right)(V_r V_c - b) = RT_r T_c.$$

But $p_c = \frac{a}{27b^2}$, $V_c = 3b$ and $T_c = \frac{8a}{27bR}$

$$\therefore \quad \left(p_v \frac{a}{27b^2} + \frac{a}{9V_r^2 b^2}\right)(3V_r b - b) = RT_r \frac{8a}{27bR}$$

or
$$\frac{a}{27b^2}\left(p_r + \frac{3}{V_r^2}\right)b(3V_r - 1) = 8T_r \frac{a}{27b}$$

or
$$\left(p_r + \frac{3}{V_r^2}\right)(3V_r - 1) = 8T_r.$$

This is the Van der Waals' *reduced equation of state.* It is independent of the constants a and b which are characteristic of a particular gas. Hence, it holds for any Van der Waals' gas.

Law of Corresponding States: It follows from the reduced equation of state that "when two of the three quantities p_r, V_r and T_r are the same for any two gases, the third quantity will also be the same for those two gases." This is known as the *law of corresponding states.*

External Work done by a Van der Waals' Gas in Isothermal Expansion: Let us consider 1 mole of a Van der Waals' gas at a temperature T K. If it expands from an initial volume V_i to a final volume V_f at constant temperature, the external work done by it can be represented as,

$$W = \int_{V_i}^{V_f} p \, dV,$$

where p is the instantaneous pressure of the gas while suffering an infinitesimal expansion dV.

The equation of state for 1 mole of the gas is,

$$\left(p+\frac{a}{V^2}\right)(V-b)=RT$$

or

$$p = \frac{RT}{V-b}-\frac{a}{V^2}$$

$$\therefore \quad W = \int_{V_i}^{V_f}\left(\frac{RT}{V-b}-\frac{a}{V^2}\right)dV$$

$$= RT\left[\log_e (V-b)\right]_{V_i}^{V_f}+a\left[\frac{1}{V}\right]_{V_i}^{V_f}$$

$$= RT\log_e\frac{V_f-b}{V_i-b}+a\left(\frac{1}{V_f}-\frac{1}{V_i}\right).$$

For μ moles of the gas this expression would be,

$$W = \mu RT\log_e\frac{V_f-\mu b}{V_i-\mu b}+a\mu^2\left(\frac{1}{V_f}-\frac{1}{V_i}\right).$$

PROBLEMS

1. ***(a) One mole of an ideal gas has a volume of 0.55 litre at 0°C. Compute its pressure.***

(b) What would be the pressure if the gas follows the Van der Waals' equation of state instead of being ideal. Take the Van der Waals' constants, a = 0.37 newton-sq metre/sq mole, b = 43 sq cm/mole and R = 8.31 joule/(mole-K)

(c) Compute the critical temperature of the gas.

Solution: (a) For 1 mole of an ideal gas, the equation of state is,

$$pV = RT.$$

Here V = 0.55 litre/mole = 550 sq cm/mole = 550 × 10^{-6} cu metre/mole; T = 273 K and R = 8.31 joule/mole-K.

$$\therefore \quad p = \frac{RT}{V} = \frac{8.31 \times 273}{550 \times 10^{-6}} = 4.125 \times 10^6 \text{ N/sq m} = 41 \text{ atoms.}$$

(b) The Van der Waals' gas equation is,

$$\left(p + \frac{a}{V^2}\right)(V - b) = RT$$

or
$$p = \frac{RT}{V-b} - \frac{a}{V^2}.$$

Here, a = 0.37 N-m⁴/mol², V = 550 × 10^{-6} m³/mol. b = 43 cm³/mol=43 × 10^{-6} m³/mol, so that $(V - b)$ = (550 – 43) × 10^{-6} = 507 × 10^{-6} m³/mol.

$$\therefore \quad p = \frac{8.31 \times 273}{507 \times 10^{-6}} - \frac{0.37}{\left(550 \times 10^{-6}\right)^2}$$

$$= 4.475 \times 10^6 - 1.223 \times 10^6$$

$$= 3.252 \times 10^6 \text{ N/m}^2 \simeq 32 \text{ atoms.}$$

(c) The critical temperature is expressed in terms of a and b as,

$$T_c = \frac{8a}{27bR}.$$

Substituting the values of a, b and R, we get,

$$T_c = \frac{8 \times \left(0.37 \text{ N-m}^4/\text{mol}^2\right)}{27 \times \left(43 \times 10^{-6} \text{ m}^3/\text{mol}\right) \times 8.31 \text{ J/mol-K}}$$

$$= 307\text{K} = 34°\text{C}.$$

2. *The Van der Waals' constants of oxygen are: a = 1.32 sq litre-atmos/mole² and b = 3.12 × 10⁻² litre/mole. The gas is stored at 300 K and has volume 1-2litre/mole. Find the pressure of the gas. R = 8.31 joule/(mole-K).*

Solution: The pressure of a gas obeying Van der Waals' equation of state is given by,

$$p = \frac{RT}{V-b} - \frac{a}{V^2}. \quad ...(i)$$

Here T = 300 K, V = 1.2 litre/mole = 1.2×10^{-3} cu metre/mole,

a = 1.32 sq litre-atoms/mole^2

= 1.32×10^{-6} metre^6-atmos/mole^2

[∵ 1 litre = 10^{-3} cu metre]

= $1.32 \times 10^{-6} \times 1.013 \times 10^5$ metre^4-newton/mole^2

[∵ 1 atmos = 1.013×10^5 newton/sq metre]

= 0.134 m^4-N/mol^2,

and b = 3.12×10^{-2} litre/mole

= 3.12×10^{-5} cu metre/mole

= 0.0312×10^{-3} cu metre/mole

so that,

$V - b$ = $(1.2 - 0.0312) \times 10^{-3} = 1.17 \times 10^{-3}$ cu m/mol.

Substituting these values in eq. (i), we have,

$$p = \frac{8.31\dfrac{\text{J}}{\text{mol-k}} \times 300\text{K}}{1.17 \times 10^{-3}\dfrac{\text{m}^3}{\text{mol}}} - \frac{0.313\dfrac{\text{m}^4-\text{N}}{\text{mol}^2}}{\left(1.2 \times 10^{-3}\dfrac{\text{m}^3}{\text{mol}}\right)^2}$$

$$= \frac{2.49 \times 10^3\,\text{J}}{1.17 \times 10^{-3}\,\text{m}^3} - \frac{0.134\text{N}}{1.44 \times 10^{-6}\,\text{m}^2}$$

$$= 2.13 \times 10^6 \frac{\text{N-m}}{\text{m}^3} - 0.093 \times 10^6 \frac{\text{N}}{\text{m}^2}$$

$$= 2.037 \times 10^6 \text{N/m}^2$$

$$= \frac{2.037 \times 10^6}{1.013 \times 10^5} = 20 \text{ atoms.}$$

3. ***Calculate the temperature at which 5 mole gas at 5 atmosphere pressure occupies a volume of 20 litre. Van der Waals' constants are: a = 1.32 litre2-atmos/mole2, b = 3.12 × 10^{-2} litre/mole and R = 8.31 joule/(mole-K).***

Solution: The Van der Waals' equation of state for the mole of gas is,

$$\left(p+\frac{a\mu^2}{V^2}\right)(V-\mu b)=\mu RT$$

so that

$$T = \frac{\left(p+\frac{a\mu^2}{V^2}\right)(V-\mu b)}{\mu R} \qquad \ldots\text{(i)}$$

Here ($\mu = 5, p = 5$ atmos $= 5 \times 1.013 \times 10^5 = 5.065 \times 10^5$ N/m², V = 20 litre = 20 × 10^{-3} cu metre, a = 1.32 litre²-atmos/mole² = 0.134 m⁴-N/mol² and b = 3.12 × 10^{-2} litre/mole = 0.0312 × 10^{-3} m³/mol (as calculated in last problem).

$$\frac{a\mu^2}{v^2} = \frac{0.134\times(5)^2}{(20\times10^{-3})^2} = 8.375\times10^3\ N/m^3$$

$$\therefore \quad p + \frac{a\mu^2}{V^2} = (5.065 \times 10^5) + (0.08375 \times 10^5)$$

$$= 5.15 \times 10^5 \text{ N/m}^2.$$

$$V - \mu b = (20 \times 10^{-3}) - (5 \times 0.0312 \times 10^{-3})$$

$$= 19.8 \times 10^{-3}\text{m}^3.$$

Substituting these values in eq. (i), we get,

$$T = \frac{(5.15 \times 10^5 \text{ N/m}^2)(19.8 \times 10^{-3} \text{ m}^3)}{5 \text{ mol} \times 8.31 \text{ J/mol-K}}$$

$$= 245\text{K} = -23°\text{C}.$$

4. ***The Van der Waals' constants for hydrogen are a = 0.245 atmos-sq litre/mole² and b = 2.67 × 10^{-2} litre/mole. Calculate the critical temperature and Boyle temperature.***

Solution: The critical temperature T_c of a gas is related to Van der Waals' constants a and b by,

$$T_c = \frac{8a}{27bR}, \quad \ldots(i)$$

where R is universal gas constant (for 1 mole). Here,

$$a = 0.245 \text{ atmos-sq litre/mole}^2$$

$$= 0.245 \times 10^{-6} \text{ atmos-metre}^6/\text{mole}^2$$

$$[\because 1 \text{ litre} = 10^{-3} \text{ cu metre}]$$

$$= 0.245 \times 10^{-6} \times 1.013 \times 10^5 \text{ newton- metre}^4/\text{mole}^2$$

$$[\because 1 \text{ atmos} = 1.013 \times 10^5 \text{ N/m}^2]$$

$$= 0.0248 \text{ N-m}^4/\text{mol}^2,$$

and $b = 2.67 \times 10^{-2}$ litre/mole

$$= 2.67 \times 10^{-5} \text{ cu metre/mole.}$$

Substituting these values of a and b, and R = 8.31 joule/(mole-K) in eq. (i), we get,

$$T_c = \frac{8 \times (0.0248 \text{ N-m}^4/\text{mol}^2)}{27 \times (2.67 \times 10^{-5} \text{ m}^3/\text{mol}) \times (8.31 \text{ J/mol-K})}$$

$$= 33\text{K} = -240°\text{C}.$$

The Boyle temperature of hydrogen is,

$$T_B = \frac{27}{8} T_c = \frac{27}{8} \times 33K$$

$$= 111 \text{ K} = -162°\text{C}.$$

5. *Calculate the critical temperature for helium; given that $a = 6.15 \times 10^{-5}$, $b = 9.95 \times 10^{-4}$; the unit of pressure being one atmosphere and the unit of volume being the volume of one mole of gas at N.T.P.*

Solution: The Van der Waals' equation of state for one mole of gas is,

$$\left(p+\frac{a}{V^2}\right)(V-b)=RT$$

In terms of given units, we have at N.T.P.

$$p = 1,\ V = 1 \text{ and } T = 273\text{K}.$$

$$\therefore \quad \left[1+\frac{(6.15\times10^{-5})}{(1)^2}\right]\{1-(9.95\times10^{-4})\}=R\times273$$

or $\quad (1.0000615)\ (1 - 0.000995) = R \times 273$

or $$R = \frac{1.0000615 \times 0.999005}{273}$$

$$= 3.66 \times 10^{-3}.$$

Now, the critical temperature, in terms of a and b is given by,

$$T_c = \frac{8a}{27bR}$$

$$= \frac{8\times(6.25\times10^{-5})}{27\times(9.95\times10^{-4})\times(3.66\times10^{-3})}$$

$$= 5.00 \text{ K} = -268°\text{C}.$$

6. ***Calculate the critical temperature of a gas for which $a = 0.00874$, $b = 0.0023$ and the gas constant is given by 273 $R = 1.00646$.***

Solution: $$T_c = \frac{8a}{27bR}$$

$$= \frac{8\times0.00874}{27\times0.0023\times(1.00646/273)}$$

$$= 305\text{K} = 32°\text{C}.$$

7. ***Calculate the Van der Waals' constants a and b for dry air; given that $T_c = 132$ K, $p_c = 37.2$ atmospheres, R per mole $= 82.07$ cm^3-atmos-K^{-1}.***

Solution: $$a = \frac{27R^2}{64}\times\frac{T_c^2}{p_c}$$

$$= \frac{27\,(82.07\text{ cm}^3\text{-atmos/mole-K})^2}{64} \times \frac{(132\text{ K})^2}{37.2\text{ atmos}}$$

$$= 1.33 \times 10^6 \text{ cm}^6\text{-atmos/mole}^2.$$

Similarly, $b = \frac{R}{8} \times \frac{T_c}{P_c}$

$$= \frac{82.07\text{ cm}^3\text{-atmos/mole-K}}{8} \times \frac{132\text{ K}}{37.2\text{ atmos}}$$

$$= 36.4 \text{ cu cm/mole}.$$

8. ***Calculate Van der waals' constants a and b for 1 kilomole helium. Given:*** $T_c = 5.3K$, $p_c = 2.2 \times 10^5$ N/m^2 ***and*** $R = 8.3 \times 10^3$ ***J/(kmol-K).***

Solution: $a = \frac{27R^2}{64} \times \left(\frac{T_c^2}{p_c}\right)$

$$= \frac{27\,(8.3 \times 10^3\text{ J/kmol-K})^2}{64} \times \frac{(5.3\text{ K})^2}{2.2 \times 10^5\text{ N/m}^2}$$

$$= 3.7 \times 10^3 \text{ N-m}^4\text{/kmol}^2.$$

$$b = \frac{R}{8} \times \left(\frac{T_c}{P_c}\right)$$

$$= \frac{8.3 \times 10^3 J/(kmol-k)}{8} \times \frac{5.3K}{2.2 \times 10^5 N/m^2}$$

$$= 0.025 \text{ m}^3\text{/kmol}.$$

9. ***The critical pressure and temperature for*** CO_2 ***are 73 atmospheres and 31°C, respectively. Calculate Van der Waals' constants.*** $R = 8.31$ ***joule/mole-K.***

Solution. $T_c = 31°\text{C} = 31 + 273 = 304 \text{ K}$

and $p_c = 73 \text{ atmos} = 73 \times (1.013 \times 10^5)$

$$= 73.9 \times 10^5 \text{ N/m}^2.$$

Now, the constants a and b are:

$$a = \frac{27R^2}{64} \times \frac{T_c^2}{p_c}$$

$$= \frac{27\,(8.31\,\text{J/mol-K})^2}{64} \times \frac{(304\text{ K})^2}{73.9 \times 10^5\text{ N/m}^2}$$

$$= 0.36\text{ N-m}^4/\text{mol}^2.$$

Again, $$b = \frac{R}{8} \times \frac{T_c}{P_c}$$

$$= \frac{8.31\text{ J/mol-K}}{8} \times \frac{304\text{ K}}{73.9 \times 10^5\text{ N/m}^2}$$

$$= 42.7 \times 10^{-6}\text{m}^3/\text{mol}$$

$$= 43\text{cm}^3/\text{mol}.$$

10. ***The critical temperature of hydrogen is – 240°C and the critical pressure is 12.8 × 10⁵ newton/sq metre. Compute the critical volume for one mole of hydrogen. Also estimate the diameter of a hydrogen molecule.***

Solution: The critical coefficient $\frac{RT_c}{p_c V_c}$ is roughly $\frac{8}{3}$ for all gases obeying Van der Waals' equation. Thus,

$$V_c = \frac{3RT_c}{8p_c}.$$

Here T_c = – 240°C = 33 K, p_c = 12.8 × 10^5 N/m² and R = 8.31 J/mol-K.

$$\therefore \qquad V_c = \frac{3 \times (8.31\text{ J/mol-K}) \times 33\text{K}}{8 \times \left(12.8 \times 10^5\text{ N/m}^2\right)}$$

$$= 80.3 \times 10^{-6}\text{ m}^3/\text{mol}.$$

The Van der Waals' constant b is given by

$$b = \frac{1}{3} V_c = 26.8 \times 10^{-6}\text{ m}^3/\text{mol}.$$

b (co-volume) is four times the volume occupied by the N molecules of the 1 mole gas. Thus, if d be the molecular diameter, we have

$$b = 4\times\frac{4}{3}\pi\left(\frac{d}{2}\right)^3\times N,$$

where N is Avogadro number (6.02×10^{23}/mole)

$$\therefore \quad d^3 = \frac{3b}{2\pi N}$$

$$= \frac{3\times\left(26.8\times10^{-6}\,\text{m}^3/\text{mol}\right)}{2\times3.14\times\left(6.02\times10^{23}\,/\text{mol}\right)}$$

$$= 21.3 \times 10^{-30} \text{ sq metre.}$$

$$\therefore \quad d = (21.3 \times 10^{-30})^{1/3}$$

$$= 2.77 \times 10^{-10} \text{ metre} = 2.77\text{Å}.$$

Boyle Temperature's Importance

At low pressure the gas occupies a large volume. Hence, the effect of the finite size of molecules is insignificant, and only the effect due to molecular attraction works. Hence, the product pV decreases as the pressure p increases. But at high pressures, the effect of finite size of molecules also comes in and soon predominates the effect of molecular attraction Hence, the product pV, after reaching a minimum, begins to increase with increase in pressure.

Above Boyle Temperature: At high temperatures the molecules of a gas are in much rapid motion and spend less time in attracting one another. Therefore, the effect of molecular attraction is small and the effect of the finite size of molecules is important at all pressures. Therefore, the product pV increases with increase in p from the very beginning.

At Boyle Temperature: At Boyle temperature the two effects counteract each other over a sufficient range of pressure so that as p increases, pV remains constant as demanded by Boyle's law.

Kinetic Theory: Explanation

According to the kinetic theory, the gases are made up of molecules of *infinitesimally small size* which *do not attract* one another. If this were the case, the gases would have obeyed the Boyle's law. The molecules of actual gases have finite size and do attract one another. These two factors are responsible for their deviation from Boyle's law.

Effect of Finite Size of Molecules: At any pressure, the observed volume V of a gas is greater than the actual free space available for the molecules to move, because the molecules of the gas themselves occupy a finite volume. Thus, the observed value of the product pV is greater that its real value.

Effect of Molecular Attraction: When the pressure of a gas increases, its volume decreases. Consequently, the molecules of the gas come closer and the attraction between them increases. This increased attraction means an additional pressure as it helps the molecules to come closer. Thus, for a given increase in pressure, the decrease in volume will be greater than what it would have been in the absence of molecular attraction. Hence, with increase in pressure p, the product pV will not remain constant but will decrease.

Favourable Points

Works of Thermodynamism

The state of a thermodynamic system of constant mass may be specified in terms of the variables p, V,T, U and S, between which two thermodynamic relations exist:

$$dQ = dU + pdV$$

and

$$dQ = TdS.$$

For a complete description of the behaviour of such systems some other relations are required. These relations are mathematically simplified when certain functions of the above variables are introduced.

There are four such functions, known as 'thermodynamic functions' described below:

Internal Energy (U): The internal energy U of a system is a thermodynamic variable. It is also considered as a thermodynamic function.

Suppose a system undergoes an infinitesimal reversible change from one equilibrium state to another.

The internal energy changes by an amount

$$dU = dQ - pdV$$

$$= TdS\text{-}pdV.$$

Taking the partial differentials of U, we get

$$\left(\frac{\partial U}{\partial S}\right)_V = T$$

and $$\left(\frac{\partial U}{\partial V}\right)_S = -p.$$

But dU is a perfect differential, so that

$$\frac{\partial}{\partial V}\left(\frac{\partial U}{\partial S}\right) = \frac{\partial}{\partial S}\left(\frac{\partial U}{\partial V}\right)$$

$$\therefore \quad \left(\frac{\partial T}{\partial V}\right)_S = -\left(\frac{\partial p}{\partial S}\right)_V \quad \ldots\text{(i)}$$

This is Maxwell's first thermodynamic relation.

Helmholtz Function (F): It is also called as "Helmholtz Free Energy" or "Thermodynamic Potential at Constant Volume". It is defined by the equation

$$\boxed{F = U - TS}$$

Since U, T and S are perfect differentials, F is also a perfect differential. When a system undergoes an infinitesimal reversible change from an initial equilibrium state to a final equilibrium state, the Helmholtz Free Energy changes by an amount given by

$$dF = dU - TdS - SdT.$$

But $dU = T\,dS - p\,dV$, as shown in the above case.

$$\therefore \qquad dF = (TdS - p\,dV) - TdS - S\,dT$$

$$= -\,SdT - pdV.$$

Taking the partial differentials of F, we have

$$\left(\frac{\partial F}{\partial T}\right)_V = -S$$

and

$$\left(\frac{\partial F}{\partial V}\right)_T = -p$$

Since dF is a perfect differential, we have

$$\frac{\partial}{\partial V}\left(\frac{\partial f}{\partial T}\right) = \frac{\partial}{\partial T}\left(\frac{\partial f}{\partial V}\right)$$

$$\left(\frac{\partial S}{\partial V}\right)_T = \left(\frac{\partial p}{\partial T}\right)_V$$

This is Maxwell's second relation.

Enthalpy or Total Heat Function (H): This is defined by

$$\boxed{H = U + pV}$$

For an infinitesimal reversible change, we get

$$dH = dU + pdV + Vdp.$$

But $dU + pdV = dU + dW = dQ$ (by first law) $= TdS$ (by second law)

$$\therefore \qquad dH = TdS + Vdp$$

The partial differentials of H are

$$\left(\frac{\partial H}{\partial S}\right)_p = T \quad \text{and} \quad \left(\frac{\partial H}{\partial p}\right)_S = V$$

Since dH is a perfect differential, we have

$$\frac{\partial}{\partial p}\left(\frac{\partial H}{\partial S}\right) = \frac{\partial}{\partial S}\left(\frac{\partial H}{\partial p}\right)$$

$$\therefore \quad \left(\frac{\partial T}{\partial p}\right)_S = \left(\frac{\partial V}{\partial S}\right)_p \qquad \text{...(iii)}$$

This is Maxwell's third relation.

Gibbs Function (G): This is also known as "Gibbs Free Energy " or "Thermodynamic Potential at Constant Pressure". It is defined by

$$\boxed{G = H - TS}$$

For an infinitesimal reversible process.

$$dG = dH - TdS - SdT$$

But $dH = TdS + Vdp$ (as shown above).

$$\therefore \quad dG = Vdp - SdT.$$

The partial differentials of G are

$$\left(\frac{\partial G}{\partial p}\right)_T = \text{V}$$

and

$$\left(\frac{\partial G}{\partial T}\right)_p = -S$$

Since dG is a perfect differential, we have

$$\frac{\partial}{\partial T}\left(\frac{\partial G}{\partial P}\right) = \frac{\partial}{\partial p}\left(\frac{\partial G}{\partial T}\right)$$

$$\therefore \quad \left(\frac{\partial V}{\partial T}\right)_P = -\left(\frac{\partial S}{\partial p}\right)_T \qquad \text{...(iv)}$$

This is Maxwell's fourth relation.

Importance: Thermodynamic functions are of practical importance in studying the conditions of equilibrium of a system. For example, the condition of equilibrium for a process in which the temperature and volume of the system remain constant (isothermal-isochoric process) may be expressed as

$$dF = 0. \qquad [\because dF = -pdV - SdT]$$

This means that out of the various states which a system can assume by isothermal-isochoric process, only those are stable in which the Helmholtz Free Energy F (= $U - TS$) is a minimum. The equation $dF = 0$ refers to a minimum value of F (and not to a maximum value). It follows from the fact that in any natural (irreversible) process the Helmholtz Free Energy can only fall because the entropy can only increase.

Similarly, the condition of equilibrium for a process in which the temperature and pressure of the system remain constant (isothermal-isobaric process) may be expressed as

$$dG = 0. \qquad [\because dG = Vdp - SdT]$$

This means that a system at constant temperature and pressure is in stable equilibrium when the Gibbs function G (the thermodynamic potential at constant pressure) is a minimum.

Enthalpy: The enthalpy H of a thermodynamic system is defined as

$$H = U + pV$$

When the system undergoes an infinitesimal process from an initial equilibrium state to a final equilibrium state, the change in enthalpy is

$$dH = dU + pdV + Vdp.$$

But $dU + pdV = dQ$ (from first law).

$\therefore$ $$dH = dQ + Vdp.$$

For an isobaric (pressure remaining constant) process, $dp - 0$, so that

$$dH = dQ.$$

If H_i and H_f represent the initial and final enthalpies of the system, then we have

$$H_f - H_i = Q.$$

Thus, *the change in enthalpy during an isobaric process is equal to the heat transferred* . This result is useful in engineering and chemistry where so many processes take place at atmospheric (that is, at constant) pressure.

Enthalpy in Porous Plug Experiment: An important property of the enthalpy function is in connection with a throttling process (porous plug experiment). In this process a gas is made to pass under a constant pressure through an insulated porous plug to a region of lower constant pressure. Let p_1 and V_1, be the initial pressure and volume and p_f and V_f the final pressure and volume of the gas.

The *net* external work done by the gas is $p_f V_f - p_i V_i$, Since there is no heat-exchange between the gas and the surroundings, this work is done at the cost of internal energy of the gas. Thus, if U_i and U_f be the initial and final internal energies, we have

$$U_i - U_f = p_f V_f - p_i V_i$$

or $$U_i + p_i V_i = U_f + p_f V_f$$

But $U + pV$ is defined as enthalpy H of the system.

$\therefore$ $$H_i = H_f$$

Thus, in a throttling process the initial and final enthalpies are equal.

Gibbs-Helmholtz's Equation

From eq. (ii), we have

$$dF = -\,dW - SdT$$

$$= -pdV - SdT.$$

Writing the partial differential of F at constant volume $(dV = 0)$; we have

$$\left(\frac{\partial F}{\partial T}\right)_V = -\,S$$

Substituting this value of S in eq. (iv), we get

$$\boxed{U = F - T\left(\frac{\partial F}{\partial T}\right)_V}$$

This is Gibbs-Helmholtz equation.

Importance: This equation does not involve the entropy S whose calculation is often difficult. Hence, it can easily be applied to study the thermodynamics of isothermal changes in a chemical system.

Suppose a chemical system changes isothermally at temperature T from state 1 to state 2. Then, from Gibbs-Helmholtz equation, we have

$$U_2 - U_1 = F_2 - F_1 - T\frac{\partial}{\partial T}(F_2 - F_1)$$

or $$\Delta U = \Delta F - T\left[\frac{\partial(\Delta F)}{\partial T}\right]_V$$

But from eq. (iii), $-\,\Delta F = W$, which is the work obtainable from the system during a reversible change, and from the first law of thermodynamics $-\,\Delta U = U_r$, which is the heat of reaction at constant volume.

$\therefore$ $$-U_r = -W - T\left[\frac{\partial}{\partial T}(-W)\right]$$

or $$W - U_r = T\left(\frac{dW}{dT}\right).$$

Thus, we can calculate, the variation of *W* with temperature, *i.e.* we can calculate the temperature at which the required amount of work would be obtained from the system.

Gibbs Function (or Gibbs Free Energy): The Gibbs function *G* of a thermodynamic system is defined as

$$G = H - TS, \qquad \text{...(i)}$$

where *H* is the enthalpy and *S* is the entropy of the system.

For an infinitesimal reversible process, the change in the Gibbs function is

$$dG = dH - TdS - SdT.$$

But $dH = T\,dS + V\,dp$

$$\therefore \qquad dG = Vdp - SdT. \qquad \text{...(ii)}$$

If during the reversible process, the temperature remains constant ($dT = 0$) and also the pressure remains constant ($dp = 0$), then, we have

$$dG = 0$$

or $$G = \text{constant}.$$

Thus, for changes at constant temperature and constant pressure (isothermal-isobaric process like fusion, sublimation, vaporisation, etc.) the Gibbs function remains constant.

If the process is irreversible (natural) the Gibbs function *G* falls (because entropy *S* increases). Hence, we may conclude that for changes at constant temperature and constant pressure, the Gibbs function falls or (in the limiting case) remains constant. It never increases.

Now, writing the partial differential of G at constant pressure ($dp = 0$), from eq. (ii), we have

$$\left(\frac{\partial G}{\partial T}\right)_p = -S$$

Making this substitution in eq. (i), we get

$$G = H + T\left(\frac{\partial G}{\partial T}\right)_p$$

PROBLEM

1. One gram of water when converted to steam at atmospheric pressure occupies a volume of 1671 cu cm. The latent heat of vaporisation at this temperature is 539 cal/g. Compute the increase in internal energy (ΔU), entropy (ΔS), enthalpy (ΔH) and the Gibbs function (ΔG) when one gram of water is evaporated at this temperature and pressure. Express all answers in the same system of units.

Solution: The heat taken in by 1 gm of water in converting to steam at 100 °C is given by

$$\Delta Q = mL = 1 \times 539 = 539 \text{ cal}$$

$$= 539 \times (4.18 \times 10^7)$$

$$= 22.53 \times 10^9 \text{ erg.}$$

The conversion takes place at *constant pressure,* and so the work done is

$$\Delta W = p dV$$

$$= (1.01 \times 10^6 \text{ dynes/sq cm})\ (1671-1) \text{ cu cm}$$

$$= 1.69 \times 10^9 \text{ erg.}$$

Therefore, the change in internal energy is

$$\Delta U = \Delta Q - \Delta W$$

$$= 22.53 \times 10^9 - 1.69 \times 10^9$$

$$= 20.84 \times 10^9 \text{ erg.}$$

The change in entropy is

$$\Delta S = \frac{\Delta Q}{T} = \frac{22.53\times10^{9}}{373} = 6.04\times10^{7} \text{ erg/K}$$

For the change in enthalpy, we write

$$H = U + pV$$

$$\Delta H = \Delta U + p\Delta V + V\Delta p$$

$$= \Delta U + p\Delta V$$

$$[\Delta p = 0, \text{ at constant pressure}]$$

$$= \Delta U + \Delta W$$

$$= \Delta Q$$

$$= 22.53 \times 10^{9} \text{ erg.}$$

The change in Gibbs function in an isothermal-isobaric process is zero

$$\Delta G = 0.$$

Significance of Isoenthalpic Curve

An isoenthalpic curve is the locus of all points representing equilibrium states of the same enthalpy of a thermodynamic system.

An isoenthalpic curve for a gas can be drawn by means of porous-plug experiment. The (initial) pressure and temperature p_i, and T_i on the high-pressure side of the plug are arbitrarily selected. The (final) pressure p_f on the other side of the plug is set at any value smaller than P_i and the temperature T_f of the gas is measured. This is done for a number of different values of p_f.

Thus, we get a number of points 1,2, 3, ... for the various (p_f, T_f) values corresponding to a point (p_i T_f). All these points represent equilibrium states of some constant mass of the gas, at which the gas has the same enthalpy. The locus of all these discrete points is the isoenthalpic curve.

Other curves corresponding to different enthalpies can be obtained by selecting different values of T_i.

The numerical value of the slope of an isoenthalpic curve at any point is the Joule-Kelvin coefficient at that point. The locus of all points at which the Joule-Kelvin coefficient is zero (*i.e.* the locus of the maxima of the curves) is the "inversion curve".

Isothermal Curve for an Ideal Gas is Isoenthalpic: By definition, the enthalpy of a thermodynamic system is

$$H = U + pV.$$

For an ideal gas $pV = RT$, so that

$$H = U + RT.$$

For an isothermal (T constant) process, the internal energy U of an ideal gas is constant, because it depends *only* on the (constant) temperature. Hence, enthalpy H is also constant. It means that a p-V curve for an ideal gas at constant temperature will be a locus of constant enthalpy, that is, it will be isoenthalpic.

Internal Energy and Free Energy: *The internal energy U of a system is the energy which it possesses due to its molecular constitution and motion.* In general, it is the sum of the kinetic energy of the molecules due to their motion and the potential energy of the molecules due to their mutual attraction.

The free energy of a system has been defined by

$$F = U - TS, \qquad \text{...(i)}$$

where T is the Kelvin temperature and S the entropy of the system. Suppose the system undergoes an infinitesimal reversible change. Then, the change in F will be given by

$$dF = dU - (TdS + SdT).$$

But $dU = dQ - dW$ (I law) and $dQ = TdS$ (II law); so that $dU = TdS - dW$.

$$\therefore \quad dF = (TdS - dW) - (TdS + SdT)$$

$$= - dW - SdT. \qquad \text{...(ii)}$$

If the change be isothermal ($dT= 0$), then

$$dF = - dW. \qquad \text{...(iii)}$$

Thus, the work done is exactly equal to the change in the free energy. This means that in a reversible isothermal change, the external work dW is done wholly at the cost of the free energy of the system. Hence, *the free energy of a system is the energy which is available for work in reversible isothermal change.* Thus, it corresponds exactly to the potential energy for mechanical system.

Rewriting eq. (i), we get

$$U = F+TS. \qquad \text{...(iv)}$$

This show that the internal energy U of a system consists of two parts: (i) the free energy F which is available for work in reversible isothermal changes; and (ii) the latent (or bound) energy $T\,S$ which is unavailable for useful work. As entropy S increases, the unavailable energy also increases and therefore the available energy decreases.

Energy to be Disbursed

Condition for Equilibrium

Let us consider two macroscopic system A and A' having energies denoted by E and E', respectively. Let $\Omega(E)$ be the number of microstates accessible to A when its energy lies between E and $E + \delta E$, and $\Omega'(E')$ the number of microstates accessible to A' when its energy lies between E' and $E' + \delta E'$. To an excellent approximation, we can assume all the $\Omega(E)$ states of A lying in the energy-interval between E and $E + \delta E$ simply to have an energy equal to E'. The same is true for A′.

Let us assume that the systems A and A' are in thermal contact, that is, they are free to exchange energy (in the form of heat). Although the energy of each system separately is then not constant, but the combined system A* = A + A′ is isolated so that its total energy E^* must remain unchanged. Therefore, neglecting any interaction energy, we can write,

$$E + E' = E^* = \text{constant}. \qquad \text{...(i)}$$

Let us now consider the situation when A and A′ have attained thermal equilibrium, that is, when the combined *isolated* system A* is in equilibrium. A* is equally likely to be found in each one of its accessible states (basic postulate of statistics).

Let Ω^*_{total} be the total number of states accessible to A*, and $\Omega^*(E)$ the number of states of A* which are such that the subsystem A has an energy equal to E. Then, the probability $P(E)$ that A has energy equal to E (*i.e.* it lies in the interval between E and $E+\delta E$), is given by,

$$P(E) = \frac{\Omega^*(E)}{\Omega^*_{total}} = C\Omega^*(E), \qquad \text{...(ii)}$$

where C $(=1/\Omega^*_{total})$ is some constant independent of E.

The number of states $\Omega^*(E)$ of A* can be expressed in terms of the number of states of A and A′. When A has energy E, it can be in any one of its $\Omega(E)$ possible states. Then, by e.q. (i), the system A′ must have energy $E'(=E^*-E)$, and it can be in any one of its $\Omega'(E') = \Omega'(E^*-E)$ possible states. Since every state of A can be combined with every state of A′ to give a different possible state of the combined system A*, the number of distinct states of A* when A has an energy E is given by the product,

$$\Omega^*(E) = \Omega(E)\,\Omega'(E^*-E).$$

Substituting this in eq. (ii), the probability that the system A has an energy E is given by,

$$P(E) = C\ \Omega(E)\,\Omega'(E^* - E). \qquad \text{...(iii)}$$

Now, according to the basic principle of statistical mechanics, the equilibrium state is one in which E, and correspondingly $(E^* - E)$ or E', have values such that the probability $P(E)$ is maximum. To investigate the condition for $P(E)$ to be maximum, it is more convenient to investigate the condition for $\log_e P(E)$ to be maximum. The eq. (iii) can be written as,

$$\log_e P(E) = \log_e C + \log_e \Omega(E) + \log_e \Omega'(E'), \quad \text{...(iv)}$$

where $E' = E^* - E$. The value of E which corresponds to the maximum of $\log_e P(E)$ is determined by the condition,

$$\frac{\partial\{\log_e P(E)\}}{\partial E} = 0 \quad \text{...(v)}$$

Using eq. (iv), the condition (v) becomes,

$$\frac{\partial}{\partial E}\log_e \Omega(E) + \frac{\partial}{\partial E}\log_e \Omega'(E') = 0$$

or

$$\frac{\partial}{\partial E}\log_e \Omega(E) + \frac{\partial}{\partial E'}\log_e \Omega'(E')\frac{\partial E'}{\partial E} = 0$$

But $\frac{\partial E'}{\partial E} = -1$ (because A and A′ *exchange* energy).

$$\therefore \boxed{\frac{\partial}{\partial E}\log_e \Omega(E) = \frac{\partial}{\partial E'}\log_e \Omega'(E').} \quad \text{...(vi)}$$

Thus, for equilibrium of the two systems, the function $\frac{\partial}{\partial E}\log_e \Omega(E)$ must remain constant. This function is denoted by βE. Thus,

$$\beta(E) = \frac{\partial}{\partial E}\log_e \Omega(E) = \frac{1}{\Omega(E)}\frac{\partial \Omega(E)}{\partial E}.$$

and

$$\beta'(E') = \frac{\partial}{\partial E'}\log_e \Omega'(E') = \frac{1}{\Omega'(E')}\frac{\partial \Omega'(E')}{\partial E'}.$$

The condition for equilibrium is therefore,

$$\boxed{\beta(E) = \beta'(E').} \quad \text{...(vii)}$$

The relation (vii) is the fundamental condition which determines the particular value of the energy of A (and the corresponding value of the energy of A') which occurs with the greatest probability $P(E)$.

Identification of β with 1/kT: According to the definition $\beta(E) = \frac{\partial}{\partial E}\log_e \Omega(E) = \frac{1}{\Omega(E)}\frac{\partial \Omega(E)}{\partial E}$, the parameter β has the dimensions of reciprocal of energy. Let us express $1/\beta$ as a multiple of some positive constant k having dimensions of energy:

$$\boxed{\frac{1}{\beta} = kT.}$$

The quantity T thus defined provides a measure of energy in units of the quantity k.

Now, the condition (vii) for thermal equilibrium is satisfied if,

$$\boxed{T = T'.}$$

Since equality of parameter T for the two systems is a condition for thermal equilibrium, the parameter T may be identified with temperature. (We know that two systems in thermal equilibrium are at the same temperature). Infact, T is called the *absolute temperature* of the system and is expressed in degrees. The constant k then has dimensions of energy/degree and is now known as Boltzmann's constant.

The eq. (vii) can now be written as,

$$\beta(E) = \beta'(E') = \frac{1}{kT} = \frac{1}{\Omega(E)}\frac{\partial \Omega(E)}{\partial E}.$$

Equal Partition vs. Energy

It states that *if a classical dynamical system is in thermal equilibrium at the absolute temperature T, then every independent quadratic term in its energy has a mean value equal to* $\frac{1}{2}kT$.

Mean Energy of a Harmonic Oscillator: Let us consider a particle of mass m performing simple harmonic oscillations in one dimension. Its energy is then given by,

$$E = \frac{p_x^2}{2m} + \frac{1}{2}\alpha x^2 \qquad \text{...(i)}$$

The first term is the kinetic energy of the particle in terms of its momentum p_x, and the second term is its potential energy, α is the force constant. This expression of energy has two quadratic terms.

Suppose that the oscillator is in equilibrium with a heat reservoir at a temperature T high enough so that we can describe the oscillator in terms of classical mechanics.

The mean energy of this oscillator is given by,

$$\overline{E} = \frac{\int E e^{-\beta E} dx\, dp_x}{\int e^{-\beta E} dx\, dp_x} \qquad \text{where } \beta = 1/kT$$

and the integrals extend over all possible values of x and p_x $(-\infty$ to $+\infty)$.

Using eq. (i), the above expression can be written as,

$$\overline{E} = \frac{\int \frac{p_x^2}{2m} e^{-\beta p_x^2/2m} dp_x}{\int e^{-\beta p_x^2/2m} dp_x} + \frac{\int \frac{1}{2}\alpha x^2 e^{-\beta \alpha x^2/2} dx}{\int e^{-\beta \alpha x^2/2} dx}$$

$$= \frac{-\frac{\partial}{\partial \beta}\left[\int e^{-\beta p_x^2/2m} dp_x\right]}{\int e^{-\beta p_x^2/2m} dp_x} + \frac{-\frac{\partial}{\partial \beta}\int e^{-\beta \alpha x^2/2} dx}{\int e^{-\beta \alpha x^2/2} dx}$$

$$= -\frac{\partial}{\partial \beta}\log_e\left[\int_{-\infty}^{+\infty} e^{-\beta p_x^2/2m} dp_x\right] - \frac{\partial}{\partial \beta}\log_e\left[\int_{-\infty}^{+\infty} e^{-\beta \alpha x^2/2} dx\right]$$

Putting $y = \sqrt{\beta} p_x$ and $z = \sqrt{\beta} x$ in the first and second integrals, respectively, we can show that,

$$\bar{E} = -\frac{\partial}{\partial\beta}\left[-\frac{1}{2}\log_e \beta + \log_e \int_{-\infty}^{+\infty} e^{-y^2/2m} dy\right]$$

$$-\frac{\partial}{\partial\beta}\left[-\frac{1}{2}\log_e \beta + \log_e \int_{-\infty}^{+\infty} e^{-\alpha z^2/2} dz\right]$$

The derivatives of $\log_e \int_{-\infty}^{+\infty} e^{-y^2/2m} dy$ and $\log_e \int_{-\infty}^{+\infty} e^{-\alpha z^2/2} dz$ are each zero because they, being definite integrals, do not contain β explicitly. Hence, we get,

$$\bar{E} = -\frac{\partial}{\partial\beta}\left[-\frac{1}{2}\log_e \beta\right] - \frac{\partial}{\partial\beta}\left[-\frac{1}{2}\log_e \beta\right]$$

$$= \frac{1}{2\beta} + \frac{1}{2\beta}$$

$$= \frac{1}{2}kT + \frac{1}{2}kT = kT.$$

This result is obvious from the principle of equipartition of energy. The energy equation (i) has two independent quadratic terms and each of them must have a mean value equal to $\frac{1}{2}kT$.

PROBLEMS

1. In a system in thermal equilibrium at absolute temperature T, two states with energy difference 4.83 x 10^{-21} joule occur with relative probability e^2. Deduce the temperature (k = 1.38 x 10^{-23} joule/K).

Solution: Using canonical distribution, the probability of finding the system in a particular state i of energy E_i is given by,

$$P_i = Ce^{-\beta E_i},$$

where C is a proportionality constant and $\beta = 1/kT$. Now, if the system has two states of energies E_1 and E_2, then,

$$P_1 = Ce^{-\beta E_1} \text{ and } P_2 = Ce^{-\beta E_2}$$

$$\therefore \quad \frac{P_1}{P_2} = \frac{e^{-\beta E_1}}{e^{-\beta E_2}} = e^{\beta(E_2 - E_1)}$$

or $$\log_e \left(\frac{P_1}{P_2} \right) = \beta(E_2 - E_1) = (E_2 - E_1)/kT$$

or $$T = \frac{(E_2 - E_1)}{k \log_e (P_1/P_2)}$$

Here $E_2 - E_1 = 4.83 \times 10^{-21}$ joule

and $\log_e (P_1/P_2) = \log_e(e^2) = 2$. Therefore,

$$T = \frac{4.83 \times 10^{-21} \text{ joule}}{(1.38 \times 10^{-23} \text{ joule/K}) \times 2} = 175\,K.$$

2. *An excited state of an atom is 1-38 eV above the ground state. Calculate the number of atoms in this excited state relative to the ground state at 16000 K. ($k = 1.38 \times 10^{-23}$ joule/K).*

Solution: Let E_1 and E_2 be the energies of the ground state and the excited state. By canonical distribution, the probabilities of finding an atom in these states are,

$$P_1 = C e^{-\beta E_1}$$

and $$P_2 = C e^{-\beta E_2},$$

where C is a constant and $\beta = \frac{1}{kT}$.

$$\therefore \quad \frac{P_2}{P_1} = \frac{e^{-\beta E_2}}{e^{-\beta E_1}} = e^{-\beta(E_2 - E_1)}$$

$$= e^{-(E_2 - E_1)/kT}$$

$$= \exp\left[-\frac{1.38 \times (1.6 \times 10^{-19}) \text{ J}}{(1.38 \times 10^{-23} \text{ J/K}) \times (16000 \text{ K})} \right]$$

$$= e^{-1}$$

$$= 0.368.$$

This is the fraction of atoms in the excited state relative to the ground state.

3. *Estimate the temperature of a gas containing hydrogen atoms at which the Balmer series lines will be observed in the absorption spectrum. The energies of the ground state and the first excited state of hydrogen atom are E_1 = - 13.6 eV and E_2 = -3.39 eV, respectively.*

(k = 1.38 x 10^{-23} joule/K = 8.62 x 10^{-5} eV/K)

***Solution*:** Balmer lines in absorption can be observed provided a significant fraction of the hydrogen atoms are initially in the first excited state.

According to Boltzmann probability distribution, the ratio of the number n_2 of atoms in the first excited state to the number n_1 of atoms in the ground state, in a large sample of gas in equilibrium at absolute temperature T, is,

$$\frac{n_2}{n_1} = \frac{e^{-E_2/kT}}{e^{-E_1/kT}}$$

$$= e^{-(E_2-E_1)/kT}$$

$$= e^{-(-3.39+13.6)eV/(8.62 \times 10^{-5} eV/k)T}$$

$$= e^{-1.18 \times 10^{-5} k/T}$$

Clearly, for a significant fraction n_2/n_1, the temperature T should be of the order of 10^{s} K.

4. *Each atom of a gas has only two states, one of energy zero and the other of energy 4.6 eV. What fraction of the atoms will be in the lower state at temperature (i) 300 K, (ii) 30,000 K. Take k = 1.38 x 10^{-23} joule/K .*

***Solution*:** The probability of finding an atom in a state i of energy E_i is given by,

$$P_i = C e^{-E_i/kT}$$

where C is a constant of proportionality.

There are only two energy states, $E_1 = 0$ and $E_2 = 4.6$ eV. The probabilities of finding the atom in these states are,

$$P_1 = Ce^{-0} = C \qquad ...(i)$$

and $$P_2 = Ce^{-4.6\,eV/kT(\text{in}\,eV)} \qquad ...(ii)$$

Since the exponential term is less than 1, the state of lower energy ($E_1 = 0$) is more probable than the state of higher energy ($E_2 = 4.6$ eV), as expected.

The constant C can be determined by the normalisation condition that the probability of finding the atom in either one of these states must be unity. That is,

$$P_1 + P_2 = C + Ce^{-4.6/kT} = 1$$

or $$C = \frac{1}{1 + e^{-4.6/kT}} \qquad ...(iii)$$

By definition, the probability P_i denotes the fraction of atoms in a given state E_i. Thus:

(i) At $T = 300$ K, the fraction of atoms in the lower state is, using (i) and (iii)

$$P_1 = C = \frac{1}{1 + \exp\left[-\frac{4.6 \times (1.6 \times 10^{-19})}{(1.38 \times 10^{-23}) \times 300}\right]}$$

[∵ 1 eV = 1.6×10^{-19} joule]

$$= \frac{1}{1 + e^{-178}}$$

$\simeq 1$ (almost all atoms). [∵ $e^{-178} \simeq 5 \times 10^{-78}$]

(ii) At $T = 30{,}000$ K, the fraction of atoms in the lower state is,

$$P_1' = C = \frac{1}{1 + \exp\left[-\frac{4.6 \times (1.6 \times 10^{-19})}{(1.38 \times 10^{-23}) \times 300}\right]}$$

$$= \frac{1}{1+e^{-1.78}}$$

$$= 0.85 \qquad [\because e^{-1.78} \simeq 0.17]$$

5. (a) *A system can take only three different energy states E_1=0, E_2 = 1.38 x 10^{-21} joule, E_3 = 2.76 x 10^{-21} joule. These states occur in 2, 5, 4 different ways, respectively. Deduce the probability that at temperature 100 K the system may be (i) in one of the microstates of energy E_3 and (ii) in ground state E_1. (k = 1.38 x 10^{-23} joule/K).*

(b) Calculate the average energy of the system at 100 K.

Solution: (a) The probability P_i of finding a system in thermal equilibrium in any one particular state i of energy E_i is given by,

$$P_i = \frac{e^{-\beta E_1}}{\sum_i e^{-\beta E_1}} \qquad \beta = \frac{1}{kT}$$

The given three states E_1, E_2, E_3 occur in 2, 5, 4 different ways. The probability distributions of these states are,

$$g_1 e^{-E_1/kT} = 2e^{-0/kT} = 2e^{-0} = 2,$$

$$g_2 e^{-E_2/kT} = 5\exp\left[-\frac{1.38 \times 10^{-21}}{1.38 \times 10^{-23} \times 100}\right] = 5e^{-1} = 5 \times 0.368$$

$$= 1.84$$

and

$$g_3 e^{-E_3/kT} = 4\exp\left[-\frac{2.76 \times 10^{-21}}{1.38 \times 10^{-23} \times 100}\right] = 4e^{-2} = 4 \times 0.315$$

$$= 0.54$$

(i) The probability of the system for being in one of the microstates of E_3 is,

$$P_3 = \frac{g_3 e^{-E_3/kT}}{\sum_{i=1}^{3} g_i e^{-E_i/kT}}$$

$$= \frac{0.54}{2+1.84+0.54} = \frac{0.54}{4.38} = 0.123.$$

(ii) The probability for the ground state is,

$$P_1 = \frac{g_1 e^{-E_1/kT}}{\sum_{i=1}^{3} g_i e^{-E_i/kT}}$$

$$= \frac{2}{2+1.84+0.54} = \frac{2}{4.38} = 0.457.$$

(b) The average energy of the system at 100 K is given by,

$$\bar{E} = \frac{\sum_i E_i g_i e^{-E_i/kT}}{\sum_i g_i e^{-E_i/kT}}$$

$$= \frac{(0 \times 2) + (1.38 \times 10^{-21} \times 1.84) + (2.76 \times 10^{-21} \times 0.54)}{2+1.84+0.54}$$

$$= \frac{0 + (2.45 \times 10^{-21}) + (1.49 \times 10^{-21})}{2+1.84+0.54}$$

$$= \frac{4.03 \times 10^{-21}}{4.38} = 0.92 \times 10^{-21} \text{ joule.}$$

6. A system can exist in three energy states $E_1 = 0$, $E_2 = 1.4 \times 10^{-14}$ erg and $E_3 = 2.8 \times 10^{-14}$ erg. These states occur in 2, 5 and 4 different ways, respectively. Find the probabilities of the system of being in these states. ($k = 1.4 \times 10^{-16}$ erg/K)

0.457, 0.420, 0.123.

Separation Work

The partition function 'Z' of a system is the sum of all the Boltzmann factors over all energy states of the system:

$$Z = \sum_i e^{-\beta E_i}.$$

Let us consider a dilute ideal gas in equilibrium at an absolute temperature T. Let us focus attention on a particular molecule of the gas and regard it as a small system in thermal contact with a heat reservoir (at temperature T) consisting of all the other molecules of the gas. The probability of finding the molecule in any one of its quantum states i, where its energy is E_i, is given by the canonical distribution,

$$P_i = \frac{e^{-\beta E_i}}{\sum_i e^{-\beta E_i}} \qquad \text{where } \beta = \frac{1}{kT}$$

Now, if a molecule is found with probability P_i in a state i of energy E_i, then its mean energy is given by,

$$\bar{E} = \sum P_i E_i = \frac{\sum_i E_i e^{-\beta E_i}}{\sum_i e^{-\beta E_i}} \qquad \text{...(i)}$$

where the summation is over all possible states i of the molecule.

Let us consider the numerator of eq. (i).

$$\sum_i E_i e^{-\beta E_i} = -\sum_i \frac{\partial}{\partial \beta}\left(e^{-\beta E_i}\right) = -\frac{\partial}{\partial \beta}\left(\sum_i e^{-\beta E_i}\right)$$

Making this substitution in eq. (i), we get,

$$\bar{E} = \frac{-\frac{\partial}{\partial \beta}\left(\sum_i e^{-\beta E_i}\right)}{\sum_i e^{-\beta E_i}}$$

Let us now define the quantity $\sum_i e^{-\beta E_i} \equiv Z$. Then, we get,

$$\bar{E} = \frac{-\frac{\partial Z}{\partial \beta}}{Z} = -\frac{1}{Z}\frac{\partial Z}{\partial \beta}$$

or

$$\bar{E} = -\frac{\partial}{\partial \beta}\log_e Z.$$

The calculation of mean energy $\bar{E}$ thus requires only the evaluation of the sum Z . The sum Z over all states of the molecule is the 'partition function' of the molecule.

Again, let us consider an assembly of molecules in equilibrium at absolute temperature *T*. The probability of finding a molecule in a particular microstate of energy E_i is

$$P_i = \frac{e^{-\beta E_i}}{\sum_i e^{-\beta E_i}}, \qquad \text{where } \beta = \frac{1}{kT}.$$

Now, the denominator $\sum_i e^{-\beta E_i}$ is the partition function Z of the molecule.

$$\therefore \qquad P_i = \frac{e^{-E_i/kT}}{Z}$$

By probability concept,

$$P_i = \frac{n_i}{N}$$

where n_i is the number of molecules having energy E_i and N is the total number of molecules.

From the last two expressions,

$$n_i = \frac{N}{Z} e^{-E_i/kT}$$

The graph between n_i and E_i will be an exponential curve.

Law of Equipartition of Energy: This law states: *If a system obeying classical statistical mechanics is in equilibrium at absolute temperature T, then every independent quadratic term in its energy expression has a mean value equal to $\frac{1}{2}kT$, where k is Boltzmann's constant.**

Let us consider a system with f degrees of freedom. Classically, this system is described by f position coordinates

$q_1, q_2, ..., q_f$ and f corresponding momenta coordinates $p_1, p_2, ..., p_f$. Its energy E is then a function of these variables, that is,

$$E = E(q_1, q_2, ..., q_f ; p_1, p_2, ..., p_f).$$

Let p_i be any particular momentum. Then, the total energy can be written in the form,

$$E = E_i (P_i) + E'(q_1, q_2, ..., q_f \;;\; p_1, p_2, ..., p_f, \text{ excluding } p_i) \quad ...(i)$$

where E_i is a function of p_i only while E' depends on all the variables *except* P_i.

Suppose that the system is in thermal equilibrium with its surroundings (heat reservoir) at an absolute temperature T. The canonical distribution for the system, being a function of all the *continuous* variables, can be written as,

$$P(q_1, q_2 ..., q_f) = \frac{e^{-\beta E(q_1, ..., p_f)}}{\int e^{-\beta E(q_1, ..., p_f)}\, dq_1, ..., dp_f},$$

where $\beta = 1/kT$. This is the probability of finding the system with its position and momenta coordinates in a range near $(q_1, ..., q_f ; p_1, ..., p_f)$. By definition, the mean value of E_i is given by,

$$\overline{E}_i = \frac{\int E_i\, e^{-\beta E(q_1, ..., p_f)}\, dq_1, ..., dp_f}{\int e^{-\beta E(q_1, ..., p_f)}\, dq_1, ..., dp_f},$$

where the integrals extend over all possible values of $q_1, ..., q_f$ and $p_1, ..., p_f$. *In* view of *eq*. (i), the last eq. may be written as

$$\overline{E}_i = \frac{\int E_i\, e^{-\beta(E_i + E')}\, dq_1, ..., dp_f}{\int e^{-\beta(E_i + E')}\, dq_1, ..., dp_f}$$

$$= \frac{\int E_i\, e^{-\beta E_i}\, dp_i \int e^{-\beta E'}\, dq_1, ..., dp_f (\text{except } dp_i)}{\int e^{-\beta E_i}\, dp_i \int e^{-\beta E'}\, dq_1, ..., dp_f (\text{except } dp_i)}$$

Such a separation of integrals is possible because E' is independent of p_i. The second integrals in numerator and denominator cancel mutually. Thus,

$$\bar{E}_i = \frac{\int E_i \, e^{-\beta E_i} \, dp_i}{\int e^{-\beta E_i} \, dp_i}$$

$$= \frac{-\frac{\partial}{\partial \beta}\left[\int e^{-\beta E_i} \, dp_i\right]}{\int e^{-\beta E_i} \, dp_i}$$

$$= -\frac{\partial}{\partial \beta} \log_e \left[\int e^{-\beta E_i} \, dp_i\right]$$

The limits of the integral may be taken from $-\infty$ to $+\infty$ as the momentum p_i can assume all possible values from $-\infty$ to $+\infty$. Thus,

$$\bar{E}_i = -\frac{\partial}{\partial \beta} \log_e \left[\int_{-\infty}^{+\infty} e^{-\beta E_i} \, dp_i\right] \qquad \text{...(ii)}$$

In order to evaluate the integral in eq. (ii) we have to assume some form for E_i as a function of p_i. Let E_i be a quadratic function of p_i as it would be if it represents a kinetic energy (we know that K.E.= $p^2/2m$). Let us write,

$$\bar{E}_i = bp_i^2,$$

where b is some constant. Then, the integral in eq. (ii) takes the form,

$$\int_{-\infty}^{+\infty} e^{-\beta E_i} \, dp_i = \int_{-\infty}^{+\infty} e^{-\beta b p_i^2} \, dp_i$$

$\sqrt{\beta}\, p_i$ so that $dy = \sqrt{\beta}\, dp_i$, we obtain,

$$\int_{-\infty}^{+\infty} e^{-\beta E_i} \, dp_i = \frac{1}{\sqrt{\beta}} \int_{-\infty}^{+\infty} e^{-by^2} \, dy$$

Making this substitution in eq. (ii), we get,

$$\bar{E}_i = -\frac{\partial}{\partial \beta} \log_e \left[\frac{1}{\sqrt{\beta}} \int_{-\infty}^{+\infty} e^{-by^2} \, dy\right]$$

$$= -\frac{\partial}{\partial \beta}\left[-\frac{1}{2}\log_e \beta + \log_e \int_{-\infty}^{+\infty} e^{-by^2}\, dy\right]$$

The derivative of $\log_e \int_{-\infty}^{+\infty} e^{-by^2}\, dy$ with respect to β is zero because it, being definite integral, does not contain β explicitly. Hence, we get,

$$\therefore \quad \overline{E}_i = -\frac{\partial}{\partial \beta}\left[-\frac{1}{2}\log_e \beta\right] = \frac{1}{2\beta}$$

or

$$\boxed{\overline{E}_i = \frac{1}{2}kT,} \qquad \left[\because \beta = \frac{1}{kT}\right]$$

which is the law of equipartition of energy.

Limitation: The law of equipartition of energy is applicable only so long the system can be described classically, usually at high temperatures.

Application: Let us apply the law to evaluate the specific heat of a monatomic ideal gas. The energy of a molecule in such a gas is simply its kinetic energy, *i.e.*,

$$E_i = \frac{(px^2 + py^2 + pz^2)}{2m} = \frac{px^2}{2m} + \frac{py^2}{2m} + \frac{pz^2}{2m}$$

By equipartition law, the mean value of each of the three terms in this expression is $\frac{1}{2}kT$, where T is absolute temperature of the gas.

$$\therefore \quad \overline{E}_i = (3/2)\, kT\,.$$

Since one mole of gas contains N_A (Avogadro's number) molecules, the mean energy of the gas (per mole) is,

$$\overline{E} = N_A\left(\frac{3}{2}kT\right) = \frac{3}{2}RT\,,$$

where R (- N_A k) is the gas constant. Therefore, the molar specific heat at constant volume is given by,

$$C_V = \left(\frac{\partial \bar{E}}{\partial T}\right)_V = \frac{2}{3}R.$$

Thermodynamism's Possibility

For a given macrostate of a system in thermal equilibrium, there may be a large number of accessible microstates. *The total number of microstates corresponding to a given macrostate is called the thermodynamic probability of that particular macrostate.*

Let us consider a system containing a large number N of molecules. A macrostate of the system can be expressed in terms of the distribution of the N molecules among various cells in the phase space. Suppose, in a particular macro- state, there are $n_1, n_2, n_3, ..., n_r$ molecules in the first, second, third, ... rth cell, respectively. The number of ways this can be done is,

$$\frac{N!}{n_1!\, n_2!\, n_3! \ldots\ldots n_r!}.$$

Each different way corresponds to a different microstate. Thus,

$$\frac{N!}{n_1!\, n_2!\, n_3! \ldots\ldots n_r!}.$$

is the thermodynamic probability of the particular macrostate.

If m molecules (say) go from the ith cell to the jth cell and m molecules come from the jth cell to the ith cell, then the macrostate of the system remains the same, but the microstate changes.

Entropy and Probability: Boltzmann's Relation: We know that a given system so alters itself that in the state of thermal equilibrium its entropy is maximum. On the other hand, statistically, the system alters in the direction of increasing probability and the equilibrium state of the system is the most probable state, *i.e.* the state of maximum thermodynamic probability. Thus, in equilibrium state, both the entropy and the thermodynamic probability are maximum. This led Boltzmann to conclude that *the entropy of a system is a function of the probability of the state of the system.* That is,

$$S = f(\Omega), \qquad \text{...(i)}$$

where S is the entropy and Ω is the thermodynamic probability of the state of the system (or the total number of microstates corresponding to the given macrostate of the system).

To find out the nature of the function, let us consider two independent systems having entropies S_1 and S_2, and probabilities Ω_1 and Ω_2. Then,

$$S_1 = f(\Omega_1) \quad \text{and} \quad S_2 = f(\Omega_2).$$

The entropy of the combined system is,

$$S = S_1 + S_2 \qquad \text{(entropy is additive)}$$

$$= f(\Omega_1) + f(\Omega_2)\,. \qquad \text{...(ii)}$$

But the probability, being multiplicative, for the combined system is $\Omega_1\,\Omega_2$. Hence, by eq. (i), the entropy of the combined system should be,

$$S = f(\Omega_1\,\Omega_2). \qquad \text{...(iii)}$$

Comparing eq. (ii) and (iii), we get,

$$f(\Omega_1\,\Omega_2) = f(\Omega_1) + f(\Omega_2). \qquad \text{...(iv)}$$

The solution of eq. (iv) determines f. To solve it, we first differentiate it partially with respect to Ω_1, and then with respect to Ω_2. This gives,

$$\Omega_2 f'(\Omega_1 \Omega_2) = f'(\Omega_1) \qquad \left[f' \textit{ is } \frac{\partial f}{\partial \Omega}\right]$$

and $$\Omega_1 f'(\Omega_1 \Omega_2) = f'(\Omega_2)$$

Dividing: $$\frac{\Omega_2}{\Omega_1} = \frac{f'(\Omega_1)}{f'(\Omega_2)}$$

or $$\Omega_1 f'(\Omega_1) = \Omega_2 f'(\Omega_2)$$

Generalising this relation, we can write,

$$\Omega f'(\Omega) = \text{constant} = k \text{ (say)}$$

or $$f'(\Omega) = \frac{k}{\Omega}.$$

Integration gives,

$$f(\Omega) = k \log_e \Omega + C,$$

where C is the integration constant. Substituting this result in eq. (i), we get,

$$S = k \log_e \Omega + C. \qquad \text{...(v)}$$

Now, according to Nernst heat theorem, the entropy of a thermodynamic system tends to zero as its temperature tends to absolute zero (as $T \to 0$; $S \to 0$). Further, near absolute zero, the thermodynamic probability $\Omega \to 1$. Thus,

$$0 = k \log_e 1 + C$$

or $$C = 0.$$

$$\therefore \quad \boxed{S = k \log_e \Omega.}$$

This is the Boltzmann's entropy-probability relation. The constant k appearing in the relation has been identified as the Boltzmann's universal constant.

Ideal Gas Equation: Suppose, a vessel of volume V_1 contains 1 mole of an ideal gas. The number of molecules in the gas is N (Avogadro number). The probability of a molecule of being found in a small part of volume V of the vessel is V/V_1. The probability of any other molecule of being found in the same small part is also V/V_1. Therefore, the probability of both the molecules of being found simultaneously in the same part is $(V/V_1)^2$, because probability is multiplicative. Hence, the probability of all the N molecules of being found in the volume V is,

$$\Omega = \left(\frac{V}{V_1}\right)^N.$$

Putting this value of Ω in Boltzmann's entropy-probability relation $S = k \log_e \Omega$, we have,

$$S = k \log_e \left(\frac{V}{V_1}\right)^N$$

or

$$S = kN(\log_e V - \log_e V_1).$$

Differentiating it with respect to V (remember that V_1 is constant), we have,

$$\frac{\partial S}{\partial V} = kN\frac{1}{V} \qquad \text{...(i)}$$

The first and second laws of thermodynamics, $dQ = dU + dW = dU + p\ dV$ and $dQ = T\ dS$, give,

$$dU + p\ dV = T\ dS.$$

Differentiating it with respect to V at constant temperature, we have,

$$\left(\frac{\partial U}{\partial V}\right)_T + p = T\left(\frac{\partial S}{\partial V}\right)_T.$$

But, for an ideal gas, $\left(\frac{\partial U}{\partial V}\right)_T = 0$ (Joule's law).

$$\therefore \left(\frac{\partial S}{\partial V}\right)_T = \frac{p}{T} \qquad \text{...(ii)}$$

Comparing eq. (i) and (ii), we get,

$$\frac{kN}{V} = \frac{p}{T}.$$

or

$$pV = k\,NT.$$

But $k\ N = R$ (universal gas constant).

$$\therefore \quad pV = RT.$$

This is ideal gas equation.

Change in Entropy of an Ideal Gas in Isothermal Expansion: Suppose, a vessel of volume V_f has μ mole of an ideal gas. The number of molecules in the gas is μN, where N is Avogadro's number.

Suppose the vessel is divided by an imaginary wall and its volume on one side of the wall is V_i. The probability of all the μN molecules of being found in the volume V_i is,

$$\Omega_i = \left(\frac{V_i}{V_f}\right)^{\mu N}.$$

If the wall be removed, then the probability of all the molecules of being found in the whole volume V_f of the vessel will be 1, that is, $\Omega_f = 1$. Therefore,

$$\frac{\Omega_i}{\Omega_f} = \left(\frac{V_i}{V_f}\right)^{\mu N}. \qquad \text{...(i)}$$

Suppose, when all the molecules are in the volume V_i then the entropy of the gas is S_i; and when all the molecules occupy the whole volume V_f, the entropy becomes S_f. Then, by the entropy-probability relation $S = k \log_e \Omega$, we have,

$$S_f - S_i = k\log_e \Omega_f - k\log_e \Omega_i = k\log_e\left(\frac{\Omega_f}{\Omega_i}\right).$$

But $\frac{\Omega_f}{\Omega_i} = \left(\frac{V_f}{V_i}\right)^{\mu N}$, by eq. (i).

$$\therefore\ S_f - S_i = k\log_e\left(\frac{V_f}{V_i}\right)^{\mu N}$$

$$= k\mu N\log_e\left(\frac{V_f}{V_i}\right)$$

$$= \mu R\log_e\left(\frac{V_f}{V_i}\right). \qquad [\because kN = R]$$

Statistical Interpretation of the Second Law of Thermodynamics: The second law of thermodynamics is concerned with the direction in which any natural process occurs in a system. According to the entropy-interpretation of the law *"any natural process in a system occurs in such a direction that results in an increase in the entropy of the system"*. (In the equilibrium state the entropy remains unchanged). Statistically, an isolated system tends to approach a state (macrostate) of large probability where the number of microstates accessible to it is larger than initially. (If the system is already in its most probable state, it remains in equilibrium). This means that the direction of natural processes is controlled by the laws of probability. Hence, statistically, the second law should be modified as:

> *A natural process occurring in a system has a large probability of producing a net increase in the entropy of the system and the surroundings.*

The probability of a decrease in entropy is very small, but not zero. Under very unusual conditions, the entropy may decrease. Such events are known as 'fluctuations'. Some examples to demonstrate the statistical nature of the second law of thermodynamics are the following:

(i) Let us consider the conduction of heat from a hot stove to a kettle of water. According to the statistical interpretation of the second law, we assert, *not* that heat *must* flow from the stove to the water, but that it is highly probable that heat will flow from the stove to the water. It is to be understood that there is always a chance, although an extremely small one, that the reverse may take place.

(ii) Suppose that a vessel containing a gas is divided into two equal compartments by a partition in which there is a small trap door. On the average, the molecules of the gas will be equally divided between the two compartments, but occasionally there may be more molecules in one compartment than in the other. Similarly, there are chances, although extremely small, when all the molecules are confined to one compartment only. According to the statistical interpretation of the second law, we would say that there is a small probability of all the molecules being confined to one compartment, but a much larger probability of an even distribution of molecules between the two compartments.

The second law tells not only the possible events, it tells which events are most probable. It does not say that the process which results in an increase of order is impossible; but it says that the processes which increase order are much less probable than the processes which increase disorder.

System in Contact with a Heat Reservoir*: *Boltzmann Canonical Distribution: In practice, we come across systems which are *not* isolated, but exchange heat with their large environment. To study such systems, we consider a small system *A′* in thermal contact with a large heat reservoir A′ at absolute temperature *T*. *A′* is so large compared to A that its temperature does not change by small exchange of energy with A . Then, at equilibrium, the probability P_i of finding the system A in any one particular state *i* of energy $E_{i,}$ is,

$$P_i = C e^{-\beta E_i},$$

where $\beta = 1/kT$ and C is constant of proportionality, independent of i. This probability distribution is known as 'Boltzmann canonical distribution.' Let us prove it.

Let $\Omega'(E')$ be the number of states accessible to A′ when its energy is equal to E' (*i.e.*, when it lies between E' and $E' + \delta E$, where δE is very small compared to the separation between the energy states of A, but large enough to contain many possible states of A′). Although the reservoir A′ can have any energy E', but the combined system A* ($\equiv$ A +A′), being isolated, should have a constant total energy E^* (say). Therefore, by energy conservation, when A′ is in its state i of energy E_i, the reservoir A′ must have an energy,

$$E' = E^* - E_i. \qquad ...(i)$$

But, when A is in this *one* definite state i, the number of states accessible to the combined system A* is simply the number of states $\Omega'(E^* - E_i)$ accessible to A′. Now, the *isolated* system A* is equally likely to be found in each one of its accessible states (basic statistical postulate). Therefore, the probability *Pi* of *A* being in the state i is proportional to the number of states accessible to A* (when A is known to be in the state i). That is,

$$P_i \propto \Omega'(E^* - E_i). \qquad ...(ii)$$

Since A is very much small than A′ (reservoir), we can write,

$$E_i << E^*$$

so that $$E' \simeq E^* \qquad \text{[by eq (i)]}$$

and we can expand $\log_e \Omega'(E^* - E_i)$ by Taylor's expansion as,

$$\log_e \Omega'(E^* - E_i) = \log_e \Omega'(E^*) - E_i \left[\frac{\partial(\log_e \Omega')}{\partial E'} \right]_{E'=E^*}$$

(neglecting higher terms)

or $$\log_e \Omega'(E^* - E_i) = \log_e \Omega'(E^*) - \beta E_i, \qquad \text{...(iii)}$$

where $\beta = \left[\frac{\partial(\log_e \Omega')}{\partial E'} \right]_{E'=E^*}$, which has been identified with $1/kT$. Since the derivative $\partial(\log_e \Omega')\partial E'$ has been evaluated at the fixed energy $E' = E^*, \beta\ (= 1/kT)$ is simply the *constant* temperature parameter of the heat reservoir A′. Hence, eq. (iii) gives,

$$\Omega'(E^* - E_i) = \Omega'(E^*) e^{-\beta E_i}.$$

Making this substitution in eq. (ii), we get,

$$P_i \propto \Omega'(E^*) e^{-\beta E_i}$$

Since $\Omega'(E^*)$ is a constant, independent of i, the last expression may be written as,

$$\boxed{P_i = C e^{-\beta E_i}.} \qquad \text{...(iv)}$$

where C is a constant independent of i. This probability distribution is 'Boltzmann canonical distribution'. The exponential factor $e^{-\beta E_i}$ is called the 'Boltzmann factor'.

The constant C in eq. (iv) can be determined by the (normalisation) condition that the total probability of *all* possible states of the system A must be unity, that is,

$$\sum_i P_i = 1$$

Now eq. (iv) gives,

$$\sum_i P_i = C\sum_i e^{-\beta E_i} = 1$$

or
$$C = \frac{1}{\sum_i e^{-\beta E_i}}$$

Making this substitution in eq. (iv), we get,

$$P_i = \frac{e^{-\beta E_i}}{\sum_i e^{-\beta E_i}}. \quad \text{...(v)}$$

Applications: The canonical distribution is very important in statistical mechanics and can be applied to discuss many cases of physical interest.

It can be used to calculate the mean value of various characteristic parameters of a system which is in thermal contact with its environment (heat-reservoir). If a parameter y of a system has a value y_i in the state i of the system, then the mean value of y is given by,

$$\bar{y} = \sum_i P_i\, y_i = \frac{\sum_i y_i\, e^{-\beta E_i}}{\sum_i e^{-\beta E_i}}.$$

where the summation is over all states i of the system.

It can thus be used to compute the mean energy or mean pressure of a gas and to investigate the magnetic properties of a substance placed in a magnetic field.

Verification Experiment

Maxwell's Speed Distribution Law in an Ideal Gas: According to kinetic theory, a gas is made up of very large number of molecules which are always in a state of motion. They are frequently colliding with one another and also with the walls of the containing vessel. Hence, their speeds and directions of motion are changing.

We may therefore conclude that at any time, *molecules of all possible speeds are present in the gas.* However, the value of the root-mean-square speed of the gas molecules remains unchanged at a given temperature.

Maxwell considered the distribution of speeds among the molecules of the gas. He made the following assumptions:

(1) The gas consists of molecules with all possible speeds between 0 and ∞.

(2) When the gas is in the steady state, its density remains, on the average, uniform throughout.

(3) Though the speeds of individual molecules are changing, but definite numbers of molecules have speeds between definite ranges.

The number of molecules in a gas having speeds in the range between c and *c + dc,* irrespective of direction, at a given temperature is given by Maxwell's speed distribution law. According to this law, this number *n(c) dc* is given by

$$\boxed{n(c)dc = 4\pi N\left(\frac{m}{2\pi kT}\right)^{3/2} e^{-mc^2/2kT} c^2 dc,} \qquad \ldots\text{(i)}$$

where *m* is the mass of a molecule, *T* is the absolute temperature of the gas and *k* is Boltzmann's constant. The total number of molecules is

$$N = \int_0^\infty n(c)dc.$$

Eq. (i) is known an Maxwell's distribution law of molecular speeds. It shows that for a given gas, the distribution depends *only* on the temperature of the gas.

Maxwell's Law of Distribution

Maxwell showed that in a gas containing *N* molecules, the number of molecules having speeds between c and c + *dc* is given by,

$$n(c)\,dc = 4\pi N\left(\frac{m}{2\pi kT}\right)^{3/2} e^{-mc^2/2kT} c^2\,dc,$$

where m is the mass of a molecule, k is the Boltzmann's constant and T is the absolute temperature of the gas. This equation is known as 'Maxwell's law of distribution of molecular speeds'.

Distribution's Curve

A plot of $n(c)$ (number of molecules per unit speed interval) against c is shown in Fig. It is known as Maxwell's distribution curve.

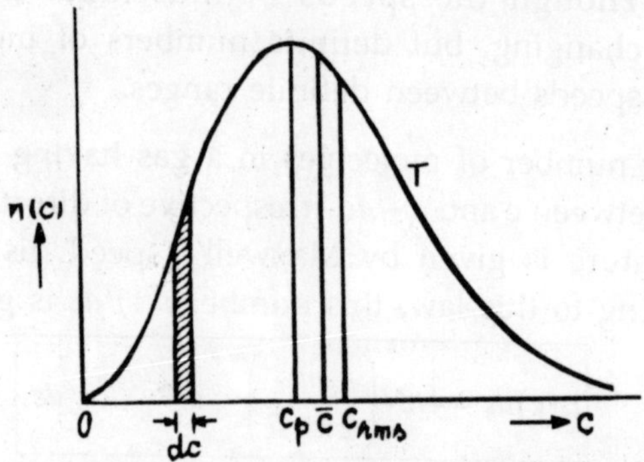

The number of molecules $n(c)\,dc$ having speeds between c and $c + dc$ is given by the area under the curve in that interval dc (shown shaded). The total area under the curve, given by the summation of $n(c)\,dc$ over all possible speeds, gives the total number of molecules:

$$\int_0^\infty n(c)\,dc = N.$$

The lower limit of the integral reflects the fact that the speed of a molecule cannot be negative.

The Distribution Curve has a Peak: For small values of c, $\frac{mc^2}{2kT} << 1$ and the exponential term $e^{-mc^2/2kT} = 1$. So, $n(c)\,dc$ (the number of molecules in speed-interval dc), being proportional to c^2, increases with increasing c. For large values of c, the exponential term dominates over c^2 and so $n(c)\,dc$, now being proportional to $e^{-mc^2/2kT}$, decreases exponentially with increasing c. This explains the appearance of a peak in the distribution curve.

The maximum number of molecules have speeds within a small range centred about the speed corresponding to the peak of the curve. This speed is called the 'most probable speed' c_p.

The Distribution Curve is Asymmetrical about its Peak: This is because the lowest possible speed of the molecules is zero but there is no limit to the highest speed a molecule can attain. Therefore, the average speed $\bar{c}$ is slightly larger than the most probable speed c_p corresponding to the peak of the curve. The root-mean-square speed c_{rms}, is still more larger $\left(c_{rms} > \bar{c} > c_p\right)$. This explains the *asymmetry* of the curve about c_p.

Influence of Temperature

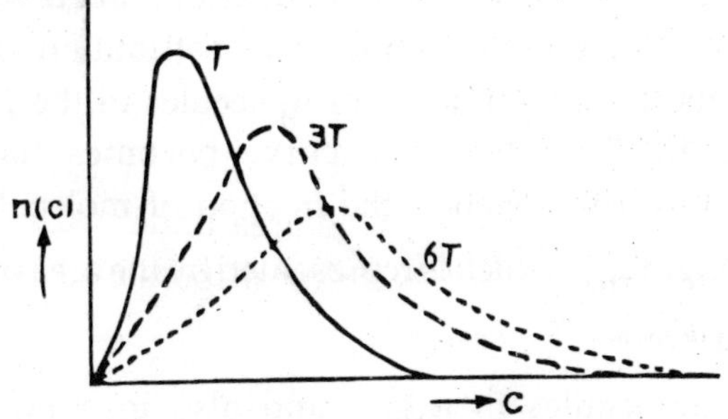

As the temperature of the gas rises, the most probable molecular speed c_p (as well as $\bar{c}$ and c_{rms}) increases in proportion to the square-root of the absolute temperature. Hence, the distribution curve broadens with rise in temperature. Since the area under the curve (which represents the total number of molecules in the gas) must remain unchanged, the curve becomes flattened as the temperature rises. This means that the number of molecules which have speeds greater than a given speed increases. In Fig. the distribution curves are shown for three different temperatures T, $3T$ and $6T$.

Effect of Molecular Mass: The most probable molecular speed c_p (as well as $\bar{c}$ and c_{rms}) is inversely proportional to the square-root of the molecular mass $\left(\propto 1/\sqrt{m}\right)$, and therefore decreases with increasing molecular mass. Thus, for example, molecular speeds in hydrogen are greater than in oxygen at the same temperature. (The average molecular *energy* is same in both gases at the same temperature.) Fig. shows the behaviour of the curve as a function of molecular mass.

As the temperature of a gas rises, the root-mean-square speed c_{rms} as well as average speed $\bar{c}$ and most probable speed c_p of the gas molecules increase (each one of them is proportional to the square-root of the kelvin temperature of the gas). For example, as the temperature rises from T to T′, the rms speed increases from its value at the point R to the value at R' (Fig.). This means that the Maxwellian distribution curve for the gas broadens. But the area under the distribution curve which represents the total number of molecules in the gas, remains unchanged. Therefore, the curve becomes flatter as the temperature rises. Hence, the fraction of molecules within a given range δc_{rms}, which is represented by the area of the shaded strip *decreases*.

The molecules in a gas, and also in a liquid, have a Maxwellian distribution of speeds. As the temperature rises, the root-mean-square speed (as well as the average speed and the most probable speed) of the molecules increases, whereas the total number of molecules remains the same. Therefore, the number of molecules which have speeds greater than some given speed increases. Hence, more molecules per second take part in chemical reactions.

According to Maxwellian law, the molecular distribution of speeds depends only upon the mass of the molecule and the temperature. The presence of other kinds of molecules in *complete equilibrium* affects neither the mass nor the temperature. Hence, the speed distribution also remains unaffected.

The distribution of speeds of molecules in a liquid resembles Maxwell's distribution (Fig.). Therefore, some of the fast molecules overcome the attraction of the surface molecules and escape (evaporation) at temperatures well below the normal boiling point. The average kinetic energy (and hence the average speed) of the remaining molecules, therefore, decreases which results in the cooling of the liquid.

The escape velocity on earth is 11.2 km/s whereas on the moon it is only 2.38 km/s. The average speed of the molecules of all atmospheric gases at the moon temperature is larger than 2.38 km/s. Hence, these molecules cannot stay at moon.

The average kinetic energy of a molecule

$$= \frac{1}{2} m \left[\frac{c_1^2 + c_2^2 + c_3^2 + \ldots c_N^2}{N} \right]$$

$$= \frac{1}{2} m\overline{c^2}$$

$$= \frac{1}{2} m (c_{rms})^2. \qquad \left[\because c_{rms} = \sqrt{\overline{c^2}} \right]$$

Thus, it is the root-mean-square speed, c_{rms}, which corresponds to a molecule having average kinetic energy.

The number of molecules in a gas is represented by the area under the distribution curve (Fig.). As the curve is non-symmetric, the area on the right side (towards increasing velocity) of the peak of the curve is slightly greater than half. We know that the peak of the curve corresponds to the most probable speed c_p. This means that more than half molecules have speeds greater than the most probable speed.

Brownian Motion's Importance

In 1827, an English botanist Brown, while examining a suspension of fine inanimate spores in water under high-power microscope, observed that the spores were constantly dancing about in water. He further found that the colloidal particles of inorganic substances also show the same type of motion in the solution. The particles move to and for rapidly and continuously, in an entirely haphazard manner. This irregular motion is called the 'Brownian motion' of the particles. It can also be observed in the smoke particles in air.

Explanation: Einstein gave a theoretical explanation of the Brownian motion and showed it to be the best evidence of the existence of thermal motion of atoms and molecules. He assumed that the particles suspended in a fluid (liquid or gas) take part in the thermal motion of the molecules of the liquid. These particles are continuously struck on all sides by the molecules of the fluid.

Since the particles are of very small size (although much larger compared with the size of the molecules), the number of molecules striking them on opposite sides do not always exactly balance. Hence, a particle always experiences an *unbalanced* force and is continuously driven about in random directions. Thus, these particles behave as 'large' molecules of the fluid.

Determination of Avogadro's Number: The colloidal particles suspended in a liquid behave like a gas which is in equilibrium under the influence of gravity. Therefore, their density varies with height in the liquid in the same way as the pressure of the atmosphere varies with altitude.

Let us consider a small element of the gas at a height h from the bottom (Fig.). Let A be the face area, and dh the thickness of the element. The mass of the element is $\rho A \, dh$ and its weight is $\rho g A \, dh$, where ρ is the density of the gas at the height h.

Let p be the (upward) pressure on the lower face and $p + dp$ (downward) that on the upper face of the element. The upward force acting on the element is pA, and the downward force is $(p + dp) A$ plus the weight of the element $\rho g A\, dh$. Since the element is in equilibrium, the resultant force on it must be zero. That is,

$$pA = (p + dp) A + \rho g A\, dh$$

or $$dp = -\rho g\, dh. \qquad \text{...(i)}$$

Now, for 1 mole of an ideal gas, we have,

$$pV = RT.$$

But $V = M/\rho$ where M is the molecular weight of the gas. If m be the mass of a gas particle and N the Avogadro's number, then $M = mN$, so that $V = mN/\rho$.

$$\therefore \quad p\frac{mN}{\rho} = RT$$

or $$p = \frac{\rho RT}{mN}.$$

Therefore, eq. (i) gives,

$$\frac{dp}{p} = -\frac{mNg}{RT} dh.$$

If n represents the number of particles per unit volume in the liquid, then $\frac{dn}{n} = \frac{dp}{p}$.

$$\therefore \quad \frac{dn}{n} = \frac{mNg}{RT} dh.$$

Suppose at the bottom, the number of particles per unit volume is n_0. That is, $n = n_0$ at $h = 0$. Then, integrating the last expression between the limits $n = n_0$ and $n = n$; $h = 0$ and $h = h$, *we have,*

$$\int_{n_0}^{n} \frac{dn}{n} = -\frac{mNg}{RT} dh.$$

or $$\log_e \frac{n}{n_0} = \frac{mNg}{RT} h$$

or $$\frac{n}{n_0} = e^{-mNgh/RT}$$

∴ $$n = e^{-mNgh/RT}.$$

If we consider the buoyancy effect of the liquid on the suspended particles, then the effective mass m of the particle would be $V(\rho - \rho')$, where V is the volume of a particle, ρ the density of particles and ρ' the density of the liquid. The last eq. gives

$$n = n_0 e^{-NV(\rho-\rho')gh/RT}$$

Perrin tested this equation and used it to determine the value of the Avogadro's number N. We have

$$\text{N} = \frac{RT}{V(\rho-\rho')gh} \log_e \frac{n}{n_0}$$

Perrin measured the number of given particles suspended at different heights in a liquid drop and determined the mass and density of the particles. He obtained a value $N = 6.8 \times 10^{23}$ particles/mole.

PROBLEMS

1. ***You are given the following group of particles. (N_i represents the number of particles with speed c_i):***

N_i	c_i **(metre/sec)**
2	10
4	20
8	30
6	40
3	50

Find out the average speed c, root-mean-square speed c_{rms}, and point out the most probable speed c_p for the entire group.

Solution: The average speed, $\bar{c}$, is the average of the speeds of all the 23 particles. Thus,

$$\bar{c} = \frac{2(1.0)+4(2.0)+8(3.0)+6(4.0)+3(5.0)}{2+4+8+6+3}$$

$$= \frac{2+4+24+24+15}{2+4+8+6+3}$$

$$= \frac{73}{23} = 3.17 \text{ metre/sec.}$$

The mean-square-speed is the average value of the squares of speeds for all the particles.

$$\overline{c^2} = \frac{2(1.0)^2+4(2.0)^2+8(3.0)^2+6(4.0)^2+3(5.0)^2}{2+4+8+6+3}$$

$$= \frac{2+16+72+96+75}{2+4+8+6+3}$$

$$= \frac{261}{23} = 11.35.$$

Therefore, the root-meant-square speed is,

$$c_{rms} = \sqrt{\overline{c^2}} = \sqrt{11.35} = 3.37 \text{ meter/sec.}$$

In the above entire group of particles the maximum number of particles (8) possess a speed of 3.0 metre/sec, hence the most probable speed is,

$$c_p = 3.0 \text{ metre/sec.}$$

2. *The speeds of 10 particles in m/s are 0, 1.0, 2.0, 3.0, 3.0, 3.0, 4.0, 4.0, 5.0, and 6.0. Find the average speed, rms speed and the most probable speed.*

Ans: 3.1 m/s, 3.5 m/s, 3.0 m/s.

3. *Calculate the rms speed and the most probable speed of the molecules of a gas whose density is 1.4 gm/litre at a pressure of 10^6 dyne/sq cm.*

Solution. From kinetic theory, the rms speed of the molecules of a gas of density ρ is given by

$$c_{rms} = \sqrt{\frac{3P}{\rho}}$$

Here $p = 10^6$ dyne/sq cm and $\rho = 1.4$ gram/litre $= 1.4 \times 10^{-3}$ gram/cu cm.

$$\therefore \qquad c_{rms} = \sqrt{\frac{3\times10^6}{1.4\times10^{-3}}} = 4.63\times10^4 \text{ cm/sec.}$$

Now, the most probable speed c_p is related to the rms speed by

$$c_p = \sqrt{\frac{2}{3}}c_{rms} = \frac{1.414}{1.732}c_{rms}$$

$$= 0.816 \times (4.63 \times 10^4)$$

$$= 3.78 \times 10^4 \text{ cm/sec.}$$

4. *Calculate the rms speed and the most probable speed of a hydrogen molecule at NTP. The Boltzmann's constant is k = 1.38 × 10⁻¹⁶ erg/(molecule-K) and Avogadro's number is N = 6 × 10²³/mole.*

Solution: The rms speed of a gas molecule of mass m at absolute temperature T is,

$$c_{rms} = \sqrt{\frac{3kT}{m}}.$$

Now, $k = 1.38 \times 10^{-16}$ erg/(molecule-K), $T = 0 + 273 = 273$ K and for hydrogen,

$$m = \frac{M(\text{molecular weight of hydrogen gas})}{N\,(\text{Avogadro's number})}$$

$$= \frac{2\text{ gram/mole}}{6\times10^{23}\text{/mole}} = \frac{1}{3}\times10^{-23}\text{ gram.}$$

$$c_{rms} = \sqrt{\frac{3\times(1.38\times10^{-16})\times273}{(1/3)\times10^{-23}}}$$

$$= \sqrt{9 \times 1.38 \times 10^7 \times 273}$$

$$= 1.84 \times 10^5 = 1.84 \times 10^5 \text{ cm/sec.}$$

The most probable speed is

$$c_p = \sqrt{\frac{2}{3}} c_{rms}$$

$$= \frac{1.414}{1.732} \times (1.84 \times 10^5) = 1.50 \times 10^5 \text{ cm/sec.}$$

5. ***Calculate the most probable speed of the molecules of a gas at 127°C. The mass of a gas molecule is 3 × 10^{-23} gm and k = 1.38 × 10^{-23} joule/K.***

Solution: The most probable speed of a gas molecule of mass m at absolute temperature T is

$$c_p = \sqrt{\frac{2kT}{m}}.$$

Given: $m = 3 \times 10^{-23}$ g = 3×10^{-26} kg, $T = 127 + 273 = 400$K and $k = 1.38 \times 10^{-23}$ J/K.

$$\therefore \quad c_p = \sqrt{\frac{2 \times (1.38 \times 10^{-23} \text{ J/K}) \times 400\text{K}}{3 \times 10^{-26} \text{ kg}}}$$

$$= 6.07 \times 10^2 \text{ m/s.}$$

6. ***Find the temperature at which the average speed of H_2 molecules will be equal to the average speed of N_2 molecules at 40°C. Molecular weights of nitrogen and hydrogen gases are 28 and 2, respectivelyf.***

Solution: The average speed $\bar{c}$ of the molecules, each of mass m, of a gas at absolute temperature T is given by

$$\bar{c} = \sqrt{\frac{8kT}{\pi m}},$$

where k is Boltzmann's constant. For $\bar{c}$ to be same for hydrogen and nitrogen, we can write

$$\bar{c} = \sqrt{\frac{8kT_H}{\pi m_H}} = \sqrt{\frac{8kT_N}{\pi m_N}}$$

or $$T_H = \frac{m_H}{m_N} \times T_N$$

Now $$\frac{m_H}{m_N} = \frac{\text{molecular weight of } H_2}{\text{molecular weight of } N_2} = \frac{2}{28}$$

and $$T_N = 40 + 273 = 313\text{K}.$$

$\therefore$ $$T_H = \frac{2}{28} \times 313 = 22.4 \text{ K} = -250.6°\text{C}.$$

7. *Calculate the root-mean-square speed of smoke particles at N.T.P. Assume the mass of a particle equal to 5 × 10-14 gram and k = 1.38 × 10^{-16} erg/(molecule-K).*

Solution: The fine smoke particles suspended in the atmosphere behave as large molecules of the atmospheric gases and have, on the average, a translational kinetic energy $\frac{3}{2}kT$, where k is Boltzmann's constant and T the kelvin temperature. Thus, if c_{rms} be the root-mean square speed, we have

$$\frac{1}{2}m(c_{rms})^2 = \frac{3}{2}kT$$

$\therefore$ $$c_{rms} = \sqrt{\frac{3kT}{m}}$$

$$= \sqrt{\frac{3 \times (1.38 \times 10^{-16}) \times 273}{5 \times 10^{-14}}} = 1.5 \text{ cm/sec.}$$

8. *A particle under Brownian motion at 27°C has a rms speed of 1 cm/sec. Find the mass of the particle. Boltzmann constant k = 1.38 × 10^{-16} erg/K.*

Solution: The translational kinetic energy of the particle is $\frac{3}{2}kT$. Thus,

$$\frac{1}{2}m(c_{rms})^2 = \frac{3}{2}kT$$

$$\therefore \quad m = \frac{3kT}{(c_{rms})^2}.$$

Here T = 27°C = 300 K and c_{rms} = 1 cm/sec.

$$\therefore \quad m = \frac{3\times(1.38\times10^{-16})\times300}{(1)^2} = 1.24\times10^{-13}\,g.$$

9. ***Particles of an inorganic substance, each having a mass of 1.55 × 10⁻¹⁴ g are suspended in a liquid at 300 K and have a root-mean-square speed of 2.8 cm/s. Compute Avogadro's number.***

Solution: The suspended particles behave like gas molecules and have an average translational kinetic energy $\frac{3}{2}kT$. Thus, if m be the mass and c_{rms} the rms velocity, then,

$$\frac{1}{2}m(c_{rms})^2 = \frac{3}{2}kT$$

$$\therefore \quad k = \frac{m}{3T}(c_{rms})^2$$

$$= \frac{1.55\times10^{-14}}{3\times300}(2.8)^2$$

$$= 1.35 \times 10^{-16} \text{ erg/(molecule-K)}.$$

Therefore, Avogadro's number is,

$$N = \frac{R}{k} = \frac{8.31\times10^{7}\text{ erg/(mole-K)}}{1.35\times10^{-16}\text{ erg/(molecule-K)}}$$

$$= 6.05\times10^{23}\text{ molecules/mole}.$$

10. *Perrin, in a water suspension of gambogo at 20°C, observed 49 particles/sq cm at the bottom and 14 particles/*

cm^2 in a layer 60 micron higher. The density of gambogo was 1.194 g/cu cm and each particle was a spherical grain of radius 0.212 micron. Find the mass of each particle, Avogadro's number, molecular weight of a particle treating it as a molecule, and the average kinetic energy of a particle, $k = 1.38 \times 10^{-16}$ erg/K, $R = 8.31 \times 10^7$ erg/(mole-K).

Solution: Mass of the particle

(radius = 0.212 μ = 0.212×10^{-4} cm)

$$m = \text{volume} \times \text{density}$$

$$= \frac{4}{3} \times 3.14 \times (0.212 \times 10^{-4})^3\text{V} \times (1.194)$$

$$= 4.77 \times 10^{-14}\text{g.}$$

The Avogadro's number is given by,

$$N = \frac{RT}{V(\rho - \rho')gh} \log_e \frac{n_0}{n},$$

where V is the volume of a particle, ρ the density of particles, ρ′ the density of the liquid, n_0 the number of particles at the bottom and n the number of particles at height h. Putting the given data, we have

$$N = \frac{(8.31 \times 10^7) \times 293 \times 2.3 \log_{10}(49/14)}{\frac{4}{3} \times 3.14 \times (0.212 \times 10^{-4})^3 \times (1.194 - 1) \times 980 \times (60 \times 10^{-4})}$$

$$= \frac{(8.31 \times 10^7) \times 293 \times 2.3 \times 0.5441}{\frac{4}{3} \times 3.14 \times (0.212 \times 10^{-4})^3 \times 0.194 \times 980 \times 60 \times 10^{-4}}$$

$$= 6.7 \times 10^{23}$$

Molecular weight of a particle is,

$$M = mN = (4.77 \times 10^{-14}) \times (6.7 \times 10^{23})$$

$$= 3.19 \times 10^{10}\text{g.}$$

The average kinetic energy of each particle is,

$$\tfrac{2}{3}kT = \tfrac{3}{2}\times\left(1.38\times10^{-16}\right)\times293$$
$$= 6.06 \times 10^{-14} \text{ erg.}$$

11. ***Find out the rms speed of colloidal particles of molecular weight 3.2 × 10⁵, if that of an hydrogen molecule is 1000 metre/sec.***

Solution:
$$\frac{c_{1rms}}{c_{2rms}} = \sqrt{\frac{M_2}{M_1}}$$

The molecular weight of hydrogen is 2.

$$\frac{c_{1rms}}{c_{2rms}} = \sqrt{\frac{2}{3.2\times10^5}} = \frac{1}{400}$$

$$\therefore \quad c_{1rms} = \frac{1}{400}\times c_{2rms}$$
$$= \frac{1}{400}\times1000$$
$$= 2.5 \text{ metre/sec.}$$

Testification of Experiment

Stem in 1926 determined experimentally the actual distribution of speeds of the molecules. The arrangement consists of a small oven *O* (Fig.) and a pair of wheels W_1 and W_2 mounted on the same shift. Each wheel has a fine slot cut in it, and the slot in W_2 is displaced by 2° relative to the slot in W_1.

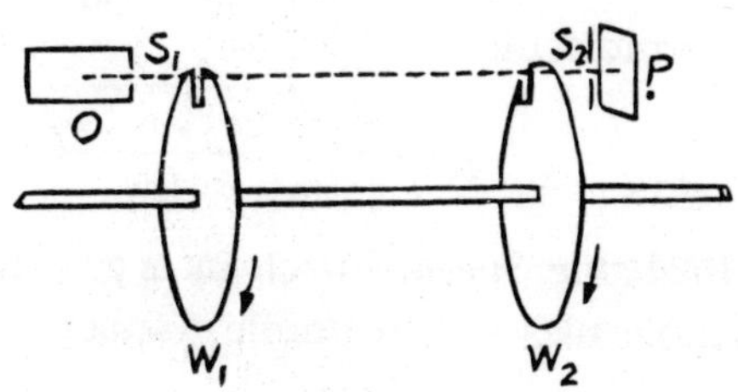

The wheels are set in *rapid* rotation. Mercury is then heated in the oven. The beam of mercury molecules issuing from O and passing through a slit S_1 falls on the slot in W_1. The molecules passing through this slot fall on the wheel W_2 but only those molecules are able to pass through the slot in W_2 whose speed is such that during the time they travel from W_1 to W_2, the wheel W_2 turns through 2°. These molecules collect on the plate P. Now, by rotating the wheels with various speeds, we collect molecules of different speeds (or speed-ranges, as the slots have finite width) at different places on the plate P. The relative intensifies of these collections can be measured by a microphotometer. From this measurement, the relative numbers of molecules lying in different speed-ranges are calculated. These numbers are found in agreement with Maxwell's distribution law.

Mean Speed: The speeds of individual molecules of a gas vary over wide range. The mean speed is the average of the speeds possessed by all the molecules. For example, if N molecules have speeds $c_1, c_2, ...,$ then the mean speed is given by

$$\overline{c} = \frac{c_1 + c_2 + c_3 + ...c_N}{N}.$$

Root-mean-square Speed: The mean of the squares of the speeds of all the molecules of the gas is called the mean-square-speed. Thus,

$$\overline{c^2} = \frac{c_1^2 + c_2^2 + c_3^2 + ...c_N^2}{N}.$$

The square-root of $\overline{c^2}$ is called the root-mean-square (rms) speed of the molecules.

$$c_{rms} \sqrt{\overline{c^2}} = \sqrt{\frac{c_1^2 + c_2^2 + c_3^2 + ...c_N^2}{N}}.$$

Most Probable Speed: Infact, in a gas there are a finite number of molecules and an infinite number of possible speeds.

We can divide the range of speeds into intervals and the probability that a molecule has a speed somewhere in a given interval has a definite value. Thus, definite number of molecules have speeds between definite ranges. It is found that at a given temperature, the largest number of molecules have speeds within a small range centred about a certain value. This value of the speed is the 'most probable speed', c_p.

Maxwell's Law of Distribution of Molecular Speeds: In an ideal gas having N molecules at absolute temperature T, the number of molecules having speeds between c and $c + dc$ is given by

$$n(c)\,dc = 4\pi N\left(\frac{m}{2\pi kT}\right)^{3/2} e^{-mc^2/2kT}c^2 dc, \qquad \text{...(i)}$$

where m is the mass of a molecule and k is Boltzmann's constant. This is the Maxwell's law of distribution of molecular *speeds*.

Average (or Mean) Speed: The product of the number of molecules in the speed-interval dc at c and the speed c is $n(c)$ dc × c. The average, or mean, speed $\overline{c}$ is the sum of all such products between the limits $c = 0$ and $c = \infty$ divided by the total number of molecules N. Thus,

$$\overline{c} = \frac{\int_0^\infty n(c)\,dc \times c}{N} \qquad \text{...(ii)}$$

Substituting the value of $n(c)$ dc from eq. (i):

$$\overline{c} = 4\pi\left(\frac{m}{2\pi kT}\right)^{3/2} \int_0^\infty e^{-mc^2/2kT}c^2 dc.$$

Let $m/2kT = b$. Then

$$\overline{c} = 4\pi\left(\frac{b}{\pi}\right)^{3/2} \int_0^\infty e^{-bc^2}c^3 dc$$

$$= 4\pi\left(\frac{b}{\pi}\right)^{3/2} \frac{1}{2b^2} \qquad \left[\because \int_0^\infty e^{-bc^2}c^3 dc = \frac{1}{2b^2}\right]$$

$$= \sqrt{\frac{4}{\pi b}}$$

$$= \sqrt{\frac{8}{\pi}\frac{kT}{m}}$$

But $\sqrt{\frac{8}{\pi}} = 1.59.$

Hence, the average speed

$$\boxed{\bar{c} = \sqrt{\frac{8}{\pi}\frac{kT}{m}} = 1.59\sqrt{\frac{kT}{m}}.} \qquad \text{...(iii)}$$

Root-mean-square (rms) Speed: Corresponding to eq. (ii), the mean value of c^2, that is, the mean-square-speed is

$$\overline{c^2} = \frac{\int_0^\infty n(c)\,dc \times c^2}{N}$$

$$= 4\pi\left(\frac{m}{2\pi kT}\right)^{3/2} \int_0^\infty e^{-mc^2/2kT} c^4 dc$$

$$= 4\pi\left(\frac{b}{\pi}\right)^{3/2} \int_0^\infty e^{-bc^2} c^4 dc$$

$$= 4\pi\left(\frac{b}{\pi}\right)^{3/2} \frac{3}{8}\sqrt{\frac{\pi}{b^5}} \quad \left[\because \int_0^\infty e^{-bc^2} c^4 dc = \frac{3}{8}\sqrt{\frac{\pi}{b^5}}\right]$$

$$= \frac{3}{2b}$$

$$= \frac{3kT}{m}.$$

Hence, the root-mean-square speed is

$$\boxed{c_{rms} = \sqrt{\overline{c^2}} = \sqrt{\frac{3kT}{m}} = 1.73\sqrt{\frac{kT}{m}}.} \qquad \text{...(iv)}$$

Most Probable Speed: The most probable speed c_p is the speed possessed by the *maximum* number of molecules. It is obtained by the condition

$$\frac{dn(c)}{dc} = 0.$$

Substituting the value of *n(c)* from eq. (i), we have

$$\frac{d}{dc}\left\{4\pi N\left(\frac{m}{2\pi kT}\right)^{3/2} e^{-mc^2/2kT}c^2\right\} = 0$$

or

$$4\pi N\left(\frac{m}{2\pi kT}\right)^{3/2}$$

$$\left\{e^{-mc^2/2kT}(2c)+e^{-mc^2/2kT}\left(-\frac{mc}{kT}\right)c^2\right\} = 0$$

or

$$2c+\left(-\frac{mc}{kT}\right)c^2 = 0$$

$$c = \sqrt{\frac{2kT}{m}}.$$

Hence, the most probable speed is

$$\boxed{c_p = \sqrt{\frac{2kT}{m}} = 1.41\sqrt{\frac{kT}{m}}} \quad \ldots(v)$$

From eq. (iii), (iv) and (v), we see that

$$\boxed{c_p < \bar{c} < c_{rms}}$$

Number of Molecules having Most Probable Speed: The number of molecules having speed c is given by

$$\frac{n(c)dc}{dc} = 4\pi N\left(\frac{m}{2\pi kT}\right)^{3/2} e^{-mc^2/2kT}c^2.$$

The most probable speed is $c = \sqrt{\frac{2kT}{m}}$. Therefore, the number ΔN (say) having most probable speed is

$$\Delta N = 4\pi N\left(\frac{m}{2\pi kT}\right)^{3/2} e^{-1}\left(\frac{2kT}{m}\right)$$

$$= 4(\pi)^{-1/2} N\left(\frac{m}{2kT}\right)^{1/2} e^{-1}$$

$$= N\left(\frac{8m}{\pi kT}\right)^{1/2} e^{-1}.$$

where e is the base of natural (or Napierian) logarithms (e = 2.72).

Maximum Probability Task

The fraction of molecules having speeds between c and $c + dc$ is $\frac{n(c)dc}{N}$. This, by definition, is the probability of a molecule for having speed between c and c + dc. Thus,

$$P(c)dc = \frac{n(c)dc}{N} = 4\pi\left(\frac{m}{2\pi kT}\right)^{3/2} e^{-mc^2/2kT} c^2 dc.$$

The probability that a molecule will have speed c, *i.e.* the probability function $P(c)$ given by

$$P(c) = 4\pi\left(\frac{m}{2\pi kT}\right)^{3/2} e^{-mc^2/2kT} c^2.$$

It is maximum for $c = \sqrt{\frac{2kT}{m}}$ (most probable speed).

$$\therefore \quad P_{max}(c) = 4\pi\left(\frac{m}{2\pi kT}\right)^{3/2} e^{-1}\left(\frac{2kT}{m}\right)$$

$$P_{max}(c) = \left(\frac{8m}{\pi kT}\right)^{1/2} e^{-1}.$$

It has dimensions of (speed).

Maxwell's Distribution Law of Molecular Speeds: The number of molecules in an ideal gas having velocities in the range c and $c + dc$ is

$$n(c)dc = 4\pi N\left(\frac{m}{2\pi kT}\right)^{3/2} e^{-mc^2/2kT}c^2dc, \qquad ...(i)$$

where N is the total number of molecules in the gas at absolute temperature T.

Momentum-wise Distribution of Molecules: The magnitude of the momentum of a molecule of mass m having velocity c is,

$$p = mc$$

so that $$c = p/m$$

and $$dc = dp/m.$$

If $n(p)\,dp$ represents the number of molecules in the momentum range p and $p + dp$, then making the above substitutions for c and dc in eq. (i), we get,

$$n(p)dp = 4\pi N\left(\frac{m}{2\pi kT}\right)^{3/2} e^{-p^2/2mkT}\left(\frac{p}{m}\right)^2\frac{dp}{m}$$

or $$\boxed{n(p)dp = 4\pi N\left(\frac{1}{2\pi mkT}\right)^{3/2} e^{-p^2/2mkT}p^2\,dp.} \qquad ...(ii)$$

Energy-wise Distribution of Molecules: The kinetic energy of a molecule of mass m and velocity c (not the speed of light) is,

$$E = \frac{1}{2}mc^2$$

so that $$c = \left(\frac{2E}{m}\right)^{1/2}$$

and $$dc = \frac{1}{2}\left(\frac{2E}{m}\right)^{-1/2}\frac{2dE}{m} = (2mE)^{-1/2}\,dE.$$

If $n(E)\,dE$ represents the number of molecules in the energy range E and $E + dE$, then making the above substitutions for c and dc in eq. (i), we get,

$$n(E)\,dE = 4\pi\left(\frac{m}{2\pi kT}\right)^{3/2} e^{-E/kT}\left(\frac{2E}{m}\right)(2mE)^{-1/2}\,dE$$

$$= \frac{4\pi N}{2\sqrt{2}}\left(\frac{m}{\pi kT}\right)^{3/2} e^{-E/kT}\,\frac{\sqrt{2}}{m^{3/2}}E^{1/2}dE$$

or $$\boxed{n(E)dE = 2\pi N\left(\frac{1}{\pi kT}\right)^{3/2} e^{-E/kT}E^{1/2}dE.} \quad \text{...(iii)}$$

It is independent of m. It means that for all gases having N molecules at temperature T, the number of molecules in the energy range E and $E + dE$ is same irrespective of their mass.

For most probable energy, we must have,

$$\frac{dn(E)}{dE} = 0$$

or $$\frac{d}{dE}\left\{2\pi N\left(\frac{1}{\pi kT}\right)^{3/2} e^{-E/kT}E^{1/2}\right\} = 0$$

or $$2\pi N\left(\frac{1}{\pi kT}\right)^{3/2}$$

$$\left\{e^{-E/kT}\left(\frac{1}{2}E^{-1/2}\right)+e^{-E/kT}\left(-\frac{1}{kT}\right)E^{1/2}\right\} = 0$$

or $$\frac{1}{2}E^{1/2}+\left(-\frac{1}{kT}\right)E^{1/2} = 0$$

or $$E = \frac{1}{2}kT$$

If E_p represents the most probable energy, then,

$$E_p = \frac{1}{2}kT$$

Note that E_p is not equal to $\frac{1}{2}mc_p^2$ where c_p $\left(= \sqrt{\frac{2kT}{m}}\right)$ is the most probable speed.

Bibliography

Anotola, S. A.: *Advanced Physics,* Longman, London, 1968.

Arun, Kumar: *Problems in Physics,* Atlantic Publishers, New Delhi, 2003.

Atherton, J. C.: *Thermal Physics,* Butterworths, London, 1987.

Atiyah, M.: *The Geometry and Physics of Knots,* Cambridge, New York, 1991.

Benjamin, W. A.: *Problems in Physics,* Academic Press, New York, 1970.

Bowling, F.: *Basic Physics,* Chapmann and Hall, London, 1976.

Caro, D. E.: *Modern Physics: An Introduction to Atomic and Nuclear Physics,* Edward Arnold, London, 1985.

Cartwright, R.: *Elements of Thermal and Statistical Physics,* Oxford University Press, London, 1966.

Clinton, C. M.: *Physical Principles of Flow in Unsaturated Porous Media,* Oxford, New York, 1994.

Crafts, A. S.: *Phloem Transport in Plants,* W.H. Freeman, San Francisco, 1973.

Daniel, Walgraif: *Introduction to Physics,* McMilian Press, Melbourne, 1969.

Douglas, J. S.: *Advanced Guide to Modern Physics,* Pelham Books, London, 1976.

Epstein, E.: *Statistical Physics,* John Wiley and Sons, New York, 1972.

Esau, K.: *Physical Geology,* John Wiley and Sons, New York, 1965.

Evans, L. T.: *Thermal Analysis of Materials,* W.A. Benjamin, California, 1975.

Feynman, R. P.: *Elementary Particles and the Laws of Physics,* Cambridge, New York, 2002.

Feynman: *The Feymman Lectures on Physics,* Add. Wesley, London, 1989.

Francia, T. G.: *The Investigation of the Physical World,* Cambridge, New York, 1981.

Ftzler M. E.: *Plant Lectins: Molecular and Biological Aspects,* Academic Press, New York, 1985.

George, H. M.: *Elements of Physical Hydrology,* Johns Hopkins, New York, 1998.

Green, J. R.: *Modern Physics,* MacMillan, London, 1911.

Halevy, A. H.: *Handbook of Physics,* C.R.C. Press, Boca Raton, 1985.

Hart, J. W. and Sabnis, D.: *Encyclopaedia of Stastical Physics,* Springer Verlag, Berlin, 1982.

Hopkins, C. R.: *Structure and Function of Cells,* W.B. Saunders, London, 1978.

Jorgensen, T. P.: *The Physics of Golf,* Springer, New York, 1999.

Karl, K. F.: *Basic Physics,* John Wiley, London, 1996.

Keith, Lockett: *Physics in the Real World,* Cambridge, New York, 1995.

Kinet, J. M. and Bemier, G.: *Thermal and Statistical Physics,* C.R.C. Press, Boca Raton, 1981.

Kirkby, E. A. and Mengel, K.: *Thermal and Statistical Physics,* Praeger Publishers, New York, 1974.

Kirkby, E. A.: *Mathematical Physics,* The University of Leeds, Leeds, 1970.

Landshoff, P.: *Essential Quantum Physics*, Cambridge, New York, 2001.

Lipson, S. G.: *Optical Physics*, Cambridge, New York, 1995.

Lui, Lam: *Introduction to Nonlinear Physics*, Springer, New York, 1997.

Main, Lain G.: *Vibrations and Waves in Physics*, Cambridge, New York, 1998.

McKee, H. S.: *University Physics*, Oxford University Press, London, 1962.

Mercer, E. I. and Galliard, T.: *Recent Advances in the Chemistry and Biochemistry of Plant Lipids*, Academic Press, New York, 1975.

Minoru, F.: *The Physics of Structural Phase Transitions*, Springer, London, 2000.

Nabapro, F. R.: *The Physics of Creep*, T & F, New York, 1995.

Negele, J. W.: *Advances in Nuclear Physics*, Plenum, London, 1994.

Palladin, I.: *Physical Metallurgy*, Arihant Publishers, Jaipur, 1988.

Peter, H.: *Physical Metallurgy*, Cambridge, New York, 1996.

Plischke, M.: *Equilibrium Statistical Physics*, World Scientific, New York, 1994.

Rabinowitch, E.: *Thermodynamics and Statistical Physics*, John Wiley, New York, 1969.

Rawn, J. D.: *Encyclopaedia of Modern Physics*, Harper and Row, New York, 1983.

Rossi, Paolo: *The Birth of Modern Science*, Blackwell, New Jersey, 2001.

Ruhland, W.: *Encyclopaedia of Thermodynamics Physics*, Springer Verlag, Berlin, 1961.

Sagdeev, Z.: *Nonlinear Space Plasma Physics*, AIP, New Jersey, 1993.

Salisbury, F. B. and Ross, C. W.: *Thermal Physics*, Woodworth Publishing, California, 1967.

Schein, L. B.: *Electro Photography and Development Physics*, Springer, London, 1992.

Sheriff, D. W. and Meidner, H.: *Water and Plants*, Blackie and Sons, Glasgow, 1976.

Slatyer, R. O.: *The Physics of Structural Phase Transitions*, Academic Press, New York, 1967.

Srivastava, H. S.: *Elements of Physics*, Rastogi Publications, Meerut, 1990.

Thimann, K. V.: *Physics and the Real World*, CRC Press , Boca Raton, Florida, 1980.

Tolbert, N. E.: *Microbodies Peroxisomes and Glyoxysomes*, Oxford University Press, London, 1971.

Tucker, G. A. and Roberts, J. A.: *Physics and Chance*, Oxford, New York, 1999.

Verner, J. E.: *Thermal Physics*, Academic Press, New York, 1985.

Vince, Pruce D.: *Elements of Stastical and Thermodynamics Physics*, McGraw Hill, London, 1975.

William, R.: *Techniques for Nuclear and Particle Physics Experiments: A How to Approach*, Springer, London, 1994.

Young, H. D.: *University Physics*, Add. Wesley, London, 1996.

Zeevart, J. A.: *Thermal Plasma and New Materials Technology*, McGraw Hill, London, 1976.

Zhukov, L.: *Modern Physics in Twenty first Century, Springer Verlag, Berlin, 1975.*

Index

A

B

C

D

E

G

H

I

K

M

N

O

P

T

V

❑❑❑